Chemistry in Aqueous and Non-Aqueous Solvents

Chemistry in Aqueous and Non-Aqueous Solvents

Y. Mido ★ S. Taguchi
M.S. Sethi ★ S.A. Iqbal

DISCOVERY PUBLISHING HOUSE
New Delhi-110 002 (INDIA)

First Published – 2001

Reprinted – 2025

ISBN: 978-81-7141-331-7

Chemistry in Aqueous and Non-Aqueous Solvents

Published by:

DISCOVERY PUBLISHING HOUSE

4383/4B, Ansari Road, Darya Ganj

New Delhi-110 002 (India)

Phone: +91-11-23279245; 23253475; 43596065

Mobile: +91 9811179893 / +91 9871656464

E-mail: discoverybooksindia@gmail.com

orderdphbooks@gmail.com

namitwasan9@gmail.com

web: www.discoverypublishinggroup.com

Printed at:

Infinity Imaging Systems

Delhi

Preface

The teaching of Chemistry at the introductory stage becomes each day a more challenging task as the subject matter becomes more diverse and more complex. These challenges have evoked a series of responses—the present set of introductory chemistry monographs is one such. The teaching of chemistry recognises a number of problems that confront those who select text books. In order to overcome these problems, this volume **"Chemistry in Aqueous and Non-Aqueous Solvents"**—one of about fifty in the Chemistry Monograph Series—is introduced. Each volume is independent of the others deals with one of Chemistry topics and constitutes and complete entity. Each volume is more comprehensive than can be possible in a single volume text. It is intended to provide a range of topics to cover most undergraduate and chemistry main courses of study. These volumes can be used to enrich the more conventional courses of study.

Suggestions for improvement are welcome and shall be gratefully acknowledged.

Authors

Contents

1
Aqueous Solution Chemistry

INTRODUCTION

Water is a versatile solvent; it dissolves '*something*' of '*everything*' and '*everything*' of '*something*'. Although many of the properties of water as a solvent and as a medium for inorganic reactions are to some extent shared by other liquids, the study of aqueous solutions still commands a special place in inorganic chemistry. This is partly because the availability of water is enormously greater than that of any other solvent; and also partly because of the ready availability of accurate physiochemical data for aqueous solutions and each of data for solutions in non-aqueous media. It is therefore reasonable to discuss inorganic chemistry in aqueous media before taking up non-aqueous medium.

Liquid water is approximately 55 molar H_2O, in which the concentration (or activity) of liquid water is conventionally taken as unity. Except for those of a few compounds that undergo extensive hydrogen bonding to water, solubilities in this solvent are usually, in mole fraction units, relatively low; even in concentrated solutions most of the molecules present are water molecules. This chapter begins with a brief survey of the physical properties of water.

STRUCTURE AND PROPERTIES OF WATER

Water is so common a substance that we often overlook its unique properties. For example, water should be a gas at room temperature, but because of hydrogen bonding, it has a boiling point of 373.15 K at 1 atm. The structure of ice and liquid water is discussed first.

Structure of Ice

To understand the behavior of water, we must first investigate the structure of ice. There are nine known crystalline forms of ice; most of them are stable only at high pressure. Ice–I, the familiar form, has been studied thoroughly. It has a density of 0.924 g ml^{-1} at 273 K and 1 atm pressure.

There is a significant difference between H_2O,

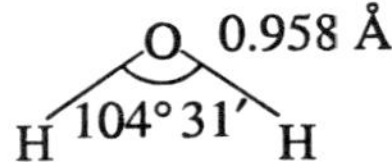

and other polar molecules, such as NH_3 and HF. In water, the number of protons about each oxygen atom that can form the positive ends of hydrogen bonds is equal to the number of lone pairs on that oxygen atom that can form the negative ends. An extensive three-dimensional network results, each oxygen atom being tetrahedrally bonded to four hydrogen atoms in two covalent bonds and two hydrogen bonds. This equality in number of protons and lone pairs is not characteristic of NH_3 and HF. Consequently, these molecules can form only rings or chains, not a three-dimensional structure.

X-ray diffraction studies do not locate the hydrogen atoms, but this can be done by a neutron diffraction investigation of solid deuterium oxide. In the common form of ice, the oxygen atoms occupy the same positions as the zinc and sulphur atoms in the wurtzite structure. The hydrogen atoms are situated slightly off the lines joining the oxygen atoms, the H—O—H angle being 105°, almost the same as in the gaseous molecule of water. The hydrogen bonding is not symmetrical, the O—H distance being 1.01 Å, a little greater than that in water vapour (0.96 Å); the overall O—H ...O distance is 2.76 Å. The wurtzite structure is a very open one (hence the low density of ice); on melting, a partial breakdown of the hydrogen-bonded network appears to occur, and some individual water molecules are accommodated in the cavities in the structure, with a consequent increase in density. Up to 277 K (4° C) (the temperature of maximum density) this effect outweighs that of thermal expansion, but above this temperature the latter effect prevails and the density falls. Even at the boiling point, however, much of the hydrogen bonding remains, as is shown by the high enthalpy and entropy of vaporisation of liquid water (Table 1.1). The strength of the hydrogen bond in ice or water has been estimated as about 20 kJ mol^{-1}.

Because of its open lattice, ice has a lower density than water, a fact of profound ecological significance. Were it not for this unique type of hydrogen bonding, ice, like most other solid substances, would be heavier than the corresponding liquid. On freezing, it would sink to the bottom of a lake or pond, causing all the water to freeze gradually. Most living organisms in the body of water would not survive. Fortunately, water reaches its maximum density at 277.15 K, which is 4 degrees above freezing. Cooling below 277.15 K decreases the density of water, allowing it to rise to the surface, where freezing occurs. An ice layer formed on the surface does not sink; just as important, it acts as a thermal insulator for the water below it.

Structure of Water

The structure of liquid water is not as well understood as that of solid ice. Interest in water is great because of its importance in chemical and biological systems, and numerous structural models have been proposed for it. Although none of the existing theories can account satisfactorily for all the observed properties of water, good progress has been made toward a detailed explanation of its structure and properties.

Bernal and Fowler in 1933 suggested that the hydrogen bonds found in ice also exist in water, although we no longer have the regular three-dimensional network of the solid. Instead, some hydrogen bonds are broken and free monomeric water molecules are formed. These monomers can occupy holes in the remaining "icelike" lattice, explaining why the density of water is greater than ice. As temperature goes up, more hydrogen bonds are broken, but at the same time the kinetic energy of molecules increases. Higher temperature increases the density of water, while elevated kinetic energy decreases the density, because each molecule will now occupy a greater volume. The net result is that a maximum density is reached at 277.15 K, above which the density decreases monatonically with increasing temperature.

According to a different theory, proposed by Pople in 1950, very few hydrogen bonds are actually broken when ice melts. Instead, the bonds are bent and distorted in liquid water. The relatively low heat of fusion for water (6.01 kJ mol^{-1}) lends some support to this model.

In 1952, Pauling showed that the structure of water may be similar to certain hydrocarbon-water complexes, such as $CH_4 \cdot 6H_2O$ and $Cl_2 \cdot 8H_2O$. Water molecules in these complexes form structures that are characteristically open like ice, but the overall arrangement is looser than that of ice. The guest molecules, CH_4 or Cl_2, are trapped inside the cavities of the icelike structure. These complexes are called *clathrate compounds* or simply *clathrates.* In the water model we would replace the guest molecule by another water molecule.

Pauling's arrangement was considered to be physically too rigid, and a closely similar *flickering-cluster model* was put forward by Frank and Wen. They assumed that the clathrates were constantly breaking up and re-forming, the mean lifetime being of the order of 10^{-10}s and further modified by Nemethy and Scheraga. Here, each water molecule can hydrogen-bond with 4, 3, 2, 1, or no water molecules. In the last case, the molecule becomes a guest in the clathrate. The manner of hydrogen bonding is cooperative: The presence of one hydrogen bond facilitates the formation of additional bonds. When one hydrogen bond is broken, the entire structure tends to disintegrate.

Some Physiochemical Properties of Water

Table 1.1 lists some important physiochemical properties of water. Several of the abnormally high values make it a unique solvent, particularly suited to the support of living systems.

Because of its high heat capacity, water acts as a good "thermostat" for proteins and nucleic acids, whose structure and function strongly depend on the temperature of their surroundings. This property has greatly influenced weather. The huge bodies of water in lakes and oceans absorb or give up large amounts of heat with only a small change in temperature. The high heat of vaporization works to reduce the loss of water by evaporation. Further, the cooling effect of evaporation helps to regulate the temperature of the water body.

Although water's heat of fusion is not very high, it is still sizable compared to other solvent. Again, this property helps to protect against freezing. Finally, we note that the high dielectric constant of water is essential for dissolving a great variety of salts, acids, and bases. The force between two charges is inversely proportional to the dielectric constant of the medium. When sodium chloride dissolves in water, the attractive force between Na^+ and Cl^- ions is greatly diminished, and the ions become hydrated through ion-dipole interaction. On the other hand, when sodium chloride dissolves in benzene (whose dielectric constant is only one-fortieth that of water), a strong attractive force keeps the ions together. Consequently, the amount of free sodium and chloride ions remaining in solution is exceedingly small.

Table 1.1 Some Physiochemical Properties of Water*

Property	Value
Melting point	273.15 K
Boiling point	373.15 K
Density of water	0.99987 g cm^{-3} (0.99987×10^3 kg m^{-3}) at 273.15 K 1.00000 g cm^{-3} (1.0000×10^3 kg m^{-3}) at 277.15 K
Density of ice	0.9167 g cm^{-3} (0.9167×10^3 kg m^{-3}) at 273.15 K
Molar heat capacity	75.3 J K^{-1} mol^{-1}
Molar heat of fusion	6.01 kJ mol^{-1}
Molar heat of vaporization	40.79 kJ mol^{-1} at 373.15 K
Dielectric constant	78.54 at 298.15 K
Dipole moment	1.82 D (6.08×10^{-30} mC)
Viscosity	0.01 P at 293.15 K (0.001 kg m^{-1} s^{-1})
Surface tension	72.75 dyn cm^{-1} at 293.15 K (0.07275 N m^{-1})
Diffusion coefficient	2.4×10^{-9} m^2 s^{-1} at 298.15 K

* When two values are given for the same quantity, the one in parentheses is in SI units.

Hydrogen bonding is mainly responsible for the very low solubility of non-polar molecules in water. The mixing of two substances is always accompanied by an increase in entropy, and on these grounds is always favoured. In the case of two liquids forming an ideal solution (i.e. one which obeys Raoult's law) there is no enthalpy of mixing; for gases or solids forming ideal solutions in liquids (e.g. nitrogen or naphthalene in hydrocarbons), the only enthalpy change is that associated with conversion of the solute into the liquid phase, and the solubility (expressed as a mole fraction) is independent of the solvent chosen. When water acts as a solvent, hydrogen bonds have to be broken. Unless the cross-interaction between the molecules or ions in a substance and water is stronger than both the interaction between the molecules or ions in the substance and the interaction between the molecules in liquid water, dissolution of the substance will be associated with an increase in the enthalpy of the system. Thus most molecular species that are readily soluble in water contain a high proportion of polar bonds (e.g. H—F, O—H, N—H) and take part in hydrogen bonding to solvent molecules. In the case of ionic species, solubility is determined by a delicate balance between lattice energy and hydration energies (resulting from ion-dipole interactions). The high dielectric constant of water plays an important part in making hydration energies of ions large and interionic forces in solution relatively small.

Water ionises according to the equation

$$2H_2O \rightleftharpoons H_3O^+ + OH^-$$

The extent of ionisation is, however, small.

On the classical (Arrhenius) theory, an acid is a compound that produces protons in aqueous solution, a base a compound that produces hydroxide ions. The role of the solvent is more clearly shown in Bronsted's theory of acids and bases, according to which an acid is a species that can act as a proton donor and a base a species that can act as a proton acceptor. For example, when hydrogen chloride (a non-electrolyte) is dissolved in water the following equilibrium is established:

$$HCl + H_2O \rightleftharpoons H_3O^+ + Cl^-$$

In this reaction water is acting as a base; hydrogen chloride is a much stronger acid than water, and the equilibrium position is far to the right. Alternatively, we may say that since water competes successfully with chloride ion for most of the protons in the system it is a much stronger base than chloride ion. In an aqueous solution of ammonia, on the other hand, water acts as an acid in the equilibrium

$$NH_3 + H_2O \rightleftharpoons NH_4^+ + OH^-$$

The difference in acid strengths between H_2O and NH_4^+, or in base strengths between NH_3 and OH^- is, however, relatively small, so that in aqueous solution NH_3 may be described as a weak base or NH_4^+ as a weak acid. Every acid is related to a conjugate base, and every base to a conjugate acid, by the relationship

$$\text{Acid} \rightleftharpoons H^+ + \text{Base}$$

Thus in the ionisation of hydrogen chloride, Cl^- is the conjugate base of HCl and HCl the conjugate acid of Cl^-. The Bronsted theory is applicable to all proton-containing systems, and we shall discuss it again (together with still wider definitions of acids and bases) in the following chapter.

In aqueous systems the H_3O^+ ion is further hydrated, and it is then often represented simply as $H^+(aq)$, a procedure we shall follow in this book; this symbol stands for H_3O^+, $H_5O_2^+$, $H_7O_3^+$, $H_9O_4^+$, and all the other hydrated protonic species that may be present in a particular solution.

CONCENTRATION UNITS

Chemists employ several different concentration units in their work, each one having its advantages as well as its limitations. The particular way of expressing the concentration of a solution is generally determined by the use of the solution. We will consider the following four concentration units: percent by weight, mole fraction, molarity, and molality.

Percent by Weight

The percent by weight of a solute in a solution is defined as

$$\text{percent by weight} = \frac{\text{weight of solute}}{\text{weight of solute} + \text{weight of solvent}} \times 100$$

$$= \frac{\text{weight of solute}}{\text{weight of solution}} \times 100$$

Mole Fraction (χ)

The mole fraction of a component A of a solution, χ_A, is defined as

$$\chi_A = \frac{\text{number of moles of component } A}{\text{number of moles of all components}}$$

The mole fraction has no units.

Molarity (M)

Molarity is defined as the number of moles of solute dissolved in 1 litre of solution that is,

$$\text{molarity} = \frac{\text{number of moles of solute}}{\text{number of litres of solution}}$$

Thus molarity has the units moles per litre. By convention, we use square brackets [ ] to represent molarity.

Molality (*m*)

Molality is defined as the number of moles of solute dissolved in 1 kg (1000 g) of solvent, that is,

$$\text{molality} = \frac{\text{number of moles of solute}}{\text{weight of solvent in kg}}$$

Thus molality has the units of moles per kg of solvent.

Having introduced these four concentration terms, we shall now compare their usefulness here. The percent by weight unit has the advantage that we do not need to know the molar mass of the solute. Furthermore, the percent by weight of a solute in a solution is independent of temperature, since it is defined in terms of weights. The mole fraction term is not normally used to express concentration of solutions. It is useful, however, for calculating partial pressures of gases. Molarity is one of the most commonly employed concentration units. The advantage of using molarity is that it is generally easier to measure the volume of a solution using precisely calibrated volumetric flasks than to weigh the solvent. Its main drawback is that it is temperature-dependent, since the volume of a solution usually increases with increasing temperature. Another drawback is that molarity does not tell us the amount of solvent present. Molality, on the other hand, is temperature-independent, since it is defined as a ratio of number of moles of solute and weight of solvent. For this reason, molality is the preferred concentration unit in studies that involve changes in temperature, as in those of the colligative properties of solution.

ELECTROLYTE SOLUTIONS

An electrolyte is a substance whose dissolving, usually in water, results in an electrically conducting solution. An electrolyte can be an acid, a base, or a salt. In this section, we will consider ionic conductance, dissociation, the thermodynamics of aqueous electrolyte solutions, and the structure and function of biological membranes.

Some Definitions

The ability of an electrolyte to conduct electricity provides us with a simple, direct means of studying ionic behaviour in solution. Let us begin by defining a few basic terms.

Ohm's Law. Ohm's law states that the current (I) flowing through a particular medium is directly proportional to the voltage or the electric potential difference (V) across the medium and inversely proportional to the resistance (R) of the medium. Thus

$$I = \frac{V}{R} \tag{1.1}$$

where I is in amperes, V in volts, and R in ohms.

Resistance. The *resistance* across a particular medium depends on the geometry of the medium; it is directly proportional to the length (l) and inversely proportional to the cross section of area (A) of the medium. Thus we write

$$R \propto \frac{l}{A}$$

$$= \rho \frac{l}{A} \tag{1.2}$$

where the proportionality constant ρ is called the *specific resistance* or *resistivity*. Since the units of R are ohms (Ω), that of l centimetres or metres, and A square centimetres or square metres, the units of ρ are Ω cm or Ω m. Resistivity is a property characteristic of the material comprising the medium.

Conductance (C). *Conductance* is the reciprocal of resistance, that is,

$$C = \frac{1}{R} = \frac{1}{\rho}\frac{A}{l} = \kappa \frac{A}{l} \tag{1.3}$$

where κ is the *specific conductance* or *conductivity*, equal to $1/\rho$. Conductivity has the units Ω^{-1} cm^{-1} or Ω^{-1} m^{-1}. In the SI system, the symbol for conductance is S (siemens).

The conductance is given by Eq. (1.3). The quantity l/A, called the *cell constant*, is the same for all solutions. Here A is the area and l the distance of separation between the electrodes. In practice, instead of measuring A and l, the cell is calibrated using a standard solution (KCl) of known κ value. Thus l/A can be calculated by measuring the conductance of the solution.

Although the specific conductance can be easily measured (from the known cell constant and the experimentally determined conductance), it is not a convenient quantity to use in the discussion of the conduction process in electrolyte solutions. Solutions of different concentrations, for example, will have very different specific conductances simply because a given volume of the different solutions will contain a different number of ions. For this reason, it is preferable to express the conductance as a molar quantity. We define the *molar conductance* (Λ_m) as

$$\Lambda_m = \frac{1000\,\kappa}{c} \tag{1.4}$$

where c is the concentration of the solution in moles per litre. The factor 1000 is necessary for Λ_m to be expressed in Ω^{-1} mol^{-1} cm^2. Another term, called the *equivalent conductance* (Λ), is also used. In this case the conductance is expressed in terms of the number of individual charges that are being carried. For example, for uni-univalent electrolyte solutions such as KCl or NaCl solution, each ion carries a unit charge, so the equivalent conductance is the same as the molar conductance. In a divalent electrolyte solution such as $MgSO_4$ each ion carries two units of charge. Here the equivalent conductance is half the molar conductance. The equivalent conductance has the units Ω^{-1} equiv^{-1} cm^2.

Conductance measurements were carried out by Kohlrausch in the nineteenth century. According to Eq. (1.4), it might appear that Λ or Λ_m would be independent of the concentration of the solution (κ is directly proportional to concentration, but κ/c should be a constant for a given substance). However, this is not the case. Instead, Kohlrausch found the following relation to hold for strong electrolytes :

$$\Lambda = \Lambda_0 - B\sqrt{c} \tag{1.5}$$

where B is a constant for a given electrolyte and Λ_0 is the equivalent conductance at infinite dilution; that is, $\Lambda \rightarrow \Lambda_0$ as $c \rightarrow 0$. Thus Λ_0 can be readily obtained by plotting Λ versus $\sqrt{c}$ and extrapolating to zero concentration. This method is unsatisfactory for weak electrolytes because of the steepness of curves at low concentrations.

Table 1.2 Equivalent Conductance at Infinite Dilution for Some Electrolytes in Water at 298 K[a]

Electrolyte	Λ_0 (Ω^{-1} equiv^{-1} cm^2)
HCl	426.16
CH_3COOH	390.71
LiCl	115.03
NaCl	126.45
AgCl	137.20
KCl	149.85
$LiNO_3$	110.14
$NaNO_3$	121.56
KNO_3	144.96
$CuSO_4$	133.62
CH_3COONa	91.00

[a] To express Λ_0 as Ω^{-1} equiv^{-1} m^2, multiply each number by 10^{-4}. Thus Λ_0 for HCl is 426.16 Ω^{-1} equiv^{-1} cm^2 or 426.16 × 10^{-4} Ω^{-1} equiv^{-1} m^2.

Table 1.2 shows the values of Λ_0 for a number of electrolytes. An interesting pattern emerges when we examine the difference in Λ_0 between two electrolytes containing a common cation or anion. For example,

$$\Lambda_0^{KCl} - \Lambda_0^{NaCl} = 23.4\ \Omega^{-1}\ \text{equiv}^{-1}\ \text{cm}^2$$

$$\Lambda_0^{KNO_3} - \Lambda_0^{NaNO_3} = 23.4\ \Omega^{-1}\ \text{equiv}^{-1}\ \text{cm}^2$$

The same difference in these two cases and similar observations led Kohlrausch to suggest the following relation :

$$\Lambda_0 = \lambda_0^+ + \lambda_0^- \tag{1.6}$$

where λ_0^+ and λ_0^- are the equivalent ionic conductances at infinite dilution. Equation (1.6) is known as *Kohlrausch's law of independent migration.* It means that equivalent conductance at infinite dilution is made up of two independent contributions from the cationic and anionic species. We can now see why the same value was obtained in the example above, since

$$\Lambda_0^{KCl} - \Lambda_0^{NaCl} = \lambda_0^{K^+} + \lambda_0^{Cl^-} - \lambda_0^{Na^+} - \lambda_0^{Cl^-} = \lambda_0^{K^+} - \lambda_0^{Na^+}$$

$$\Lambda_0^{KNO_3} - \Lambda_0^{NaNO_3} = \lambda_0^{K^+} + \lambda_0^{NO_3^-} - \lambda_0^{Na^+} - \lambda_0^{NO_3^-} = \lambda_0^{K^+} - \lambda_0^{Na^+}$$

Degree of Dissociation

At a certain concentration, an electrolyte may only be partially dissociated. At infinite dilution, any electrolyte, weak or strong, is completely dissociated. In 1887, Arrhenius suggested that the *degree of dissociation* (α) of an electrolyte can be calculated by the simple relation

$$\alpha = \frac{\Lambda}{\Lambda_0} \tag{1.7}$$

where Λ is the equivalent conductance at a particular concentration to which α refers. Equation (1.7) was used by Ostwald to measure the equilibrium constant of dissociation. Consider the dissociation of a weak acid HA of concentration c (mol litre^{-1}):

$$\underset{c(1-\alpha)}{HA} \rightleftharpoons \underset{c\alpha}{H^+} + \underset{c\alpha}{A^-}$$

By definition, the equilibrium constant is given by

$$K = \frac{[H^+][A^-]}{[HA]} = \frac{c^2\alpha^2}{c(1-\alpha)}$$

Using the expression for α in Eq. (1.7), we obtain

$$K = \frac{c\Lambda^2}{\Lambda_0(\Lambda_0 - \Lambda)} \tag{1.8}$$

Equation (1.8) can be rearranged to give

$$\frac{1}{\Lambda} = \frac{1}{K\Lambda_0^2}(\Lambda c) + \frac{1}{\Lambda_0} \tag{1.9}$$

Thus K can be obtained either directly from Eq. (1.8) or more accurately from Eq. (1.9) by plotting $1/\Lambda$ versus Λc. Of course, Λ_0 must be known if only Eq. (1.8) is employed.

Applications of Conductance Measurements

Accurate conductance measurements are easy to make and have many different applications. Two examples are described.

Acid–Base Titration. As mentioned earlier, the conductances for H^+ and OH^- are considerably higher than those for other cations and anions. By following the conductance of a HCl solution as a function of NaOH solution added. Initially, the conductance of the solution falls, since H^+ ions are replaced by Na^+ ions, which have a lower ionic conductance. This trend continues until the equivalence point is reached. Beyond this point, the conductance begins to rise, as a result of the excess OH^- ions present. If the acid is a weak electrolyte, say acetic acid, the slope of the first part of the curve is much less steep—the conductance actually increases right from the beginning—and there is more uncertainty in determining the equivalent point.

Solubility Determination. Suppose that we are interested in the solubility (mol litre^{-1}), and solubility product of AgCl in water at 298 K. From Eq. (1.4), we write

$$\Lambda = \frac{1000\kappa}{c} = \frac{1000\kappa}{S}$$

where S is the solubility in mol litre^{-1}. Since we are dealing with an insoluble salt, the concentration of the solution is low, so that we can assume that $\Lambda \simeq \Lambda_0$. Thus

$$S = \frac{1000\kappa}{\Lambda} \simeq \frac{1000\kappa}{\Lambda_0}$$

Experimentally, the specific conductance of a saturated AgCl solution is found to be $1.86 \times 10^{-6}\ \Omega^{-1}\ \text{cm}^{-1}$. However, since water is a weak electrolyte, we must take out the contribution due to water itself (κ for water is $6.0 \times 10^{-8}\ \Omega^{-1}\ \text{cm}^{-1}$). Thus

$$\kappa(\text{AgCl}) = 1.86 \times 10^{-6} - 6.0 \times 10^{-8} = 1.8 \times 10^{-6}\ \Omega^{-1}\ \text{cm}^{-1}$$

From Table 1.2, we find that $\Lambda_0 = 137.2\ \Omega^{-1}\ \text{equiv}^{-1}\ \text{cm}^2$ for AgCl. Finally, we have

$$S = \frac{1000\ \text{cm}^3\ \text{litre}^{-1} \times 1.8 \times 10^{-6}\ \Omega^{-1}\ \text{cm}^{-1}}{137.2\ \Omega^{-1}\ \text{equiv}^{-1}\ \text{cm}^2} = 1.3 \times 10^{-5}\ \text{equiv litre}^{-1}$$

$$= 1.3 \times 10^{-5}\ \text{mol litre}^{-1}$$

Since AgCl is a 1:1 electrolyte, the number of equivalents per litre is the same as the number of moles per litre. The solubility product K_{sp} for AgCl is give by

$$K_{sp} = [Ag^+][Cl^{-1}] = (1.3 \times 10^{-5})(1.3 \times 10^{-5})$$
$$= 1.7 \times 10^{-10}$$

IONS IN AQUEOUS SOLUTION

When an electrolyte (MX) is dissolved in water, two questions are relevant. First, how do the ions interact with each other? Second, how do the ions interact with the solvent molecules? Let us examine these questions.

The dissolution process can be represented by

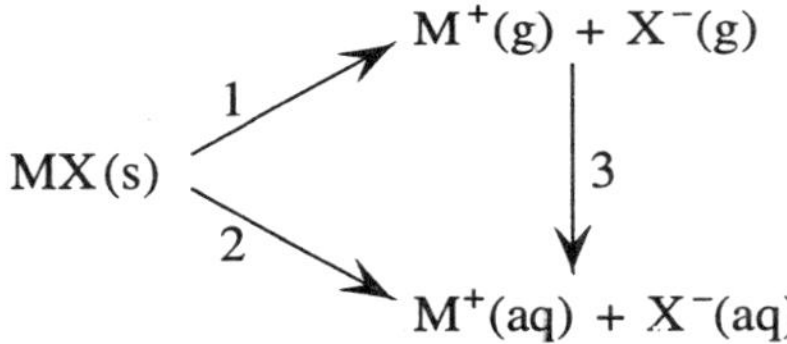

The enthalpy change for 1 corresponds to the energy required to separate the ions from the crystal lattice to infinite distance. This energy is called the *lattice energy* (U).

The enthalpy change for 2 is the heat of solution, ΔH_{soln}, the heat absorbed or released when solid MX dissolves in a large amount of water.

The heat of hydration, ΔH_{hydr}, for 3 is given by Hess's law:

$$\Delta H_{hydr} = \Delta H_{soln} - U$$

The quantity ΔH_{soln} is experimentally measurable; U can either be estimated if the structure of the lattice is known or directly measured.

For NaCl, we have $U = 765$ kJ mol^{-1} from calculation and $U = 778$ kJ mol^{-1} from experiment. Further, $\Delta H_{soln} = 3.8$ kJ mol^{-1}, so that

$$\Delta H_{hydr} = 3.8 - 778 = -774 \text{ kJ mol}^{-1}$$

Thus the hydration of Na^+ and Cl^- ions by water releases a large amount of heat.

The heat of hydration obtained above comes from both ions together. It is often desirable to know the value of individual ions. The value can be determined as follows. The hydration energy for the process

$$H^+ (g) \longrightarrow H^+ (aq)$$

has been reliably estimated to be about 1089 kJ mol^{-1}. Using this value as a starting point, it is possible to calculate the ΔH_{hydr} values for individual anions such as F^-, Cl^-, Br^-, and I^- (from data on HF, HCl, HBr, and HI), which in turn enable us to obtain ΔH_{hydr} values for Li^+, Na^+, K^+, and other cations (from data on alkali metal halides). The other quantity of interest is the entropy of hydration, ΔS_{hydr}. Table 1.3 summarizes the ΔH_{hydr} and ΔS_{hydr} for a number of ions.

Table 1.3 Thermodynamic Values for the Hydration of Gaseous Ions at 298.15 K

Ion	$-\Delta H^\circ_{hydr}$ (kJ mol^{-1})	$-\Delta S^\circ_{hydr}$ (J mol^{-1} K^{-1})	Ionic Radius (Å)
H^+	1089	109	—
Li^+	506	119	0.60
Na^+	398	87	0.95
K^+	314	52	1.33
Ag^+	468	92	1.26
Mg^{2+}	1908	268	0.65
Ca^{2+}	1577	209	0.99
Ba^{2+}	1288	159	1.35
Mn^{2+}	1832	243	0.80
Fe^{2+}	1908	272	0.76
Cu^{2+}	2092	259	0.72
Fe^{3+}	4355	460	0.64
F^-	506	151	1.36
Cl^-	377	98	1.81
Br^-	343	83	1.95
I^-	297	60	2.16

Generally, smaller ions have larger heats of hydration. A small ion contains a more concentrated charge, leading to a greater electrostatic interaction between the ion and the polar water molecules. The negative values for the entropy of hydration indicate that there is an ordered arrangement of H_2O molecules surrounding each individual ion. Because a different number of water molecules surround each type of ion, we speak of the *hydration number* of an ion. This number is directly proportional to the charge and inversely proportional to the size of the ion. It should be noted that water in the "hydration sphere" and bulk water molecules have different properties, which can be distinguished by spectroscopic techniques such as nuclear magnetic resonance. There is a dynamic equilibrium between the two types of molecules. Depending on the ion, the mean lifetime of a H_2O molecule in the hydration sphere can vary tremendously. For example, consider the mean lifetime of H_2O in the hydration sphere for the following ions: Br^- (10^{-11} s), Na^+ (10^{-9} s), Cu^{2+} (10^{-7} s), Fe^{3+} (10^{-5} s), Al^{3+} (7 s), and Cr^{3+} (1.5×10^5 s or 42 h).

The ion–dipole interaction between dissolved ions and water molecules can affect a number of bulk properties of water. Small and/or multicharged ions such as Li^+, Na^+, Mg^{2+}, Al^{3+}, Er^{3+}, OH^-, and F^- exert high electric fields. These ions, therefore, can polarize the water molecules, producing additional order beyond the first hydration layer. This interaction leads to an increase in the solutions' viscosity. On the other hand, large monovalent ions such as K^+, Rb^+, Cs^+, NH_4^+, Cl^-, NO_3^-, and ClO_4^- have diffuse surface charges and hence weak electric fields; these ions are unable to polarize water molecules beyond the first layer of hydration. Consequently, the viscosities of solutions containing these ions are often *lower* than that of pure water.

The fact that ions are hydrated in solution means that their effective radii can be appreciably greater than their crystal or ionic radii. For example, the radii of the hydrated Li^+, Na^+, and K^+ ions are estimated to be 3.66 Å, 2.80 Å, and 1.87 Å, respectively, although the ionic radii actually increase from Li^+ to K^+. We might expect that the mobility of an ion is inversely proportional to its hydrated radius. Although the proton is very small and we expect it to be strongly hydrated, it has a high ionic mobility as a result of the rapid movement of the H^+ ions via hydrogen bonds.

A survey of the values in Table 1.3 reveals several features of interest. Highly charged ions have much more negative enthalpies and entropies of hydration than ions which carry a single charge. The more negative enthalpy term is readily understood in terms of simple electrostatic attraction; the more negative entropy of hydration may be thought of as arising from the introduction of more order into the system when ions of high charge orient more water molecules in fixed configurations around them, an effect which far outweighs the increase in translational entropy when the ions go into solution. For ions of a given charge, some dependence of enthalpy and entropy of hydration on size (in the crystallographic sense) may be noted: smaller ions have more negative values of both ΔH°_H and ΔS°_H; the variation in ΔH°_H, however, outweighs that in $T\Delta S^\circ_H$, and ΔG°_H is much more negative for smaller ions of the same charge or more highly charged ions of the same size. It is, in fact, found empirically that for a cation of charge ze and crystallographic radius r_+ ångstrom units, ΔG°_H is roughly proportional to $z^2/(r_+ + 0.8)$; for anions ΔG°_H is roughly proportional to z^2/r_-.

The entropies of monatomic ions in solids or in the gaseous state are found to vary only slightly from one species to another, and it is therefore reasonable to discuss the solubilities of ionic salts in terms of differences between lattice energies and free energies of hydration of gaseous ions. These differences are small differences between large quantities; furthermore, change in ionic size along a series of related compounds affects both quantities, so that changes in them tend to cancel. Nevertheless, some helpful conclusions can be drawn. The lattice energy of a series of salts of similar structure is proportional to $1/(r_+ + r_-)$, and hence there will be only a small variation in this quantity if $r_- >> r_+$; if this condition holds, however, the free energy of hydration of the cation will be very much more negative than that of the anion for all values of r_+, and hence the sum of the free energies of hydration of the ions will be very nearly the same as the free energy of hydration of the cation alone and will be proportional to $1/(r_+ + 0.8)$. Now if, along a series of salts, the lattice energy remains nearly constant whilst the total free energy of hydration of the ions becomes less negative, solubility will decrease. This is the case with the alkali metal perchlorates and hexafluorophosphates, and similar considerations also apply to the chloroplatinates(IV) and hexanitrocobaltates(III). It may be noted that for a very small anion, such as fluoride, the variation in solubilities is reversed, with lithium fluoride being the least soluble; the minimum solubility among the chlorides occurs at potassium chloride.

All of the foregoing discussion relates to compounds for which the ionic model is valid. Many marked variations in solubilities have, of course, quite different origins. Thus among the silver halides there is a substantial increase in the extra non-electrostatic contribution to the lattice energy as we go along the series AgF, AgCl, AgBr, AgI; this progressively stabilises the solid more with respect to the ions in solution. Hence although Ag^+ is of about the same crystallographic size as Na^+ in the fluorides, the overall patterns of solubilities among the halides of the two elements are quite different.

It is generally accepted that the interaction of a simple cation such as Na^+ and the solvent is mainly ion-dipole in nature, the water molecules nearest to the ion having their oxygen atoms adjacent to it; for an anion the orientation of the water molecules is reversed. Some support for the validity of these suggestions is provided by X-ray diffraction studies of concentrated aqueous solutions of sodium and potassium hydroxides, in which the cations appear to be surrounded tetrahedrally by four oxygen atoms.

To determine hydration numbers of ions tracer methods and electronic and nuclear magnetic resonance spectroscopy are used which provide reliable information about coordination number and stereochemistry.

We now turn to the other question: How do ions interact with one another? To maintain electrical neutrality in solution, an anion must be in the vicinity of a cation, and vice versa. Depending on the proximity of these two ions, we can think of them either as "free" ions or as "ion pairs". Each free ion is surrounded by at least one, and perhaps several, layers of water molecules. In an ion pair, the cation and anion are close to each other, with little or no solvent molecules between them. Generally, free ions and ion pairs are thermodynamically distinguishable species, having quite different chemical reactivities. For dilute 1:1 aqueous electrolyte solutions, ions are believed to be in the free-ion form. In higher valent electrolytes such as $CaCl_2$ or Na_2SO_4, the formation of ion pairs is indicated by conductance measurements, for a neutral ion pair cannot conduct electricity.

Two opposing factors determine whether we have free ions or ion pairs in solution: the potential energy of attraction between the cation and anion and the kinetic or thermal energy, of the order of kT, for individual ions. In addition, we know that ion-pair formation is favoured in mixed solvents such as water–dioxane or water–alcohol. Water has a high *dielectric constant* (ε), so the electrostatic force between two ions of opposite charge is considerably reduced, as shown by the equation

$$F = \frac{Q_1 Q_2}{\varepsilon r^2} \tag{1.10}$$

Ions would then prefer to be in the free form. Adding an organic solvent decreases the dielectric constant of the solution, resulting in a greater attractive force and hence a larger amount of ion pairs. Table 1.4 lists the dielectric constant of a number of solvents. It should be kept in mind that ε always decreases with increasing temperature. For example, at 343 K the dielectric constant of water is reduced to about 64.

Table 1.4 Dielectric Constant of Some Pure Liquids at 298.15 K

Liquid	Dielectric Constant, ε^a
H_2SO_4	101
H_2O	78.54
$(CH_3)_2SO$ (dimethylsulfoxide)	49
$C_3H_8O_3$ (glycerol)	42.5
CH_3NO_2 (nitromethane)	38.6
$HOCH_2CH_2OH$ (ethylene glycol)	37.7
CH_3CN (acetonitrile)	36.2
CH_3OH	32.6
C_2H_5OH	24.3
CH_3COCH_3	20.7
CH_3COOH	6.2
C_6H_6	4.6
$C_2H_5OC_2H_5$	4.3
CS_2	2.6

[a] The dielectric constant is a dimensionless quantity.

THE IONISATION OF ACIDS IN AQUEOUS SOLUTION

Consider the aqueous equilibrium

$$HX \rightleftharpoons H^+ + X^-$$

The factors which influence the degree of ionisation are apparent from examination of the following cycle :

$$\begin{array}{ccccc}
HX_{(g)} & \xrightarrow{(2)} & H_{(g)} & + & X_{(g)} \\
\uparrow (1) & & \downarrow (3) & & \downarrow (4) \\
 & & H^+_{(g)} & & X^-_{(g)} \\
 & & \downarrow (5) & & \downarrow (6) \\
HX_{(aq)} & \rightarrow & H^+_{(aq)} & + & X^-_{(aq)}
\end{array}$$

Steps (3) and (5) are common to all the halides,

Step (2) is the dissociation of the hydrogen halide into atoms,

Steps (4) the capture of an electron by the gaseous halogen atom,

and (6) the hydration of the gaseous halide ion; for each of these steps ΔH° and ΔS° are known. Step (1), however, which is the reverse of the dissolution of the gaseous hydrogen halide in water to form an unionised solution, presents more difficulty. Since hydrogen fluoride is a weak acid in dilute aqueous solution, ΔH° and ΔS° can be obtained directly for this compound, but for the other hydrogen halides values for step (1) have to be estimated from somewhat unsatisfactory comparisons with the noble gases and the methyl halides.

We will summarise the conclusions regarding the weakness of hydrogen fluoride ($pK_a = 3.2$) relative to the other hydrogen halides ($pK_a = -7, -9$ and -10 for HCl, HBr and HI respectively). Hydrogen fluoride has a much more negative free energy of solution in water (owing to hydrogen bonding to the solvent) than the other halides; the bond in it is much stronger; and these factors outweigh the small irregular variation in electron affinity among the halogens and the more negative free energy of hydration of the fluoride ion. It will be noticed that the discussion of this thermodynamic cycle does not involve electronegativity or the inductive effect, less precise concepts than the variables we have taken into consideration; but it will also be noticed that, since four different factors enter into the determination of the relative ionisation constants, this superficially simple problem is really rather a difficult one. Similar cycles may be devised for the series H_2S, H_2Se and H_2Te and NH_3, PH_3, AsH_3 and SbH_3, in both of which ionisation constants increase with increase in atomic number; for these, however, more data have to be estimated. It is, nevertheless, clear that the decrease in bond strength with increasing atomic number plays an important part in accounting for what is often thought to be the puzzling observation that as, down a periodic table group of elements, the element becomes more metallic in character, its hydride becomes more acidic. If we consider successive ionisations of an acid like H_2S, in which both protons are attached to the same atom, it is always found that $K_2 \ll K_1$; in this instance $pK_1 = 7$, $pK_2 = 14$.

The largest group of oxo-acids are the organic carboxylic acids, many of which have pK_a values in the range 0–5 (e.g. HCOOH, 3.8; CH_3COOH, 4.8; $ClCH_2COOH$, 2.9; $Cl_2CHCOOH$, 1.3; CCl_3COOH, 0.7). It is common practice in organic chemistry to attribute these variations to the inductive effects of CH_3 and Cl, electron-repelling and electron-attracting substituents respectively; but inductive effects are not independently measurable, and since we know very little about the electron affinities of RCOO radials or the hydration energies of $RCOO^-$ ions, a quantitative treatment of the ionisation of carboxylic acids is impossible. It is, however, interesting to note that studies of the temperature-dependence of pK_a values show that for the ionisation of HCOOH, CH_3COOH and CCl_3COOH, ΔH° is in each case very close to zero, and that the variation in pK_a therefore arises from a variation in the entropy of ionisation, which becomes less negative along the series CH_3COOH, HCOOH, CCl_3COOH; apparently withdrawal of electrons from the carboxylate end of the anion results in less orientation of surroundings water molecules.

For inorganic oxo-acids, in which many different elements in different oxidation states are found, again no adequate thermodynamic treatment is possible, though there are certain useful empirical generalisations about pK_a values. The best known of these is Bell's rule that for an acid of formula, $EO_n(OH)_m$ (where E is any element) the first ionisation constant is given approximately by

$$\mathrm{p}K_a = 8 - 5n$$

Typical values are Cl(OH), 7.2 and $B(OH)_3$, 9.2; ClO(OH), 2.0 and NO(OH), 3.3; $ClO_2(OH)$, −1 and $NO_2(OH)$, −1.4; and $ClO_3(OH)$, −10. For polybasic acids it is often found that successive pK_a values differ by about 4 or 5. This increase in acid strength with increase in the number of oxygen atoms attached to the central atom is generally attributed to the greater possibility of delocalisation of negative charge on to the oxygen atoms, and this seems to be the best comment on the variation in pK_a values that can be made at the present time. It is worth noting that three acids which appear to be out of line (telluric(VI) acid, periodic acid and carbonic acid) are not really exceptions to Bell's rule; the first two are not H_2TeO_4 and HIO_4 but $Te(OH)_6$ and $IO(OH)_5$, and the last has a true pK_a value of 3.9. In an aqueous solution of carbon dioxide most of the solute is present as CO_2 rather than as H_2CO_3, but the pK_a value is usually evaluated as 6.0 on the assumption that it is all present as H_2CO_3 or the HCO_3^- ion.

COMPLEX FORMATION

The formation of complexes in aqueous solution may be recognised by a number of different methods, of which the classical test of modification of chemical properties is only one, and a somewhat unreliable one at that: all reactions have equilibrium constants, and chemical tests are often only investigations of relative values of equilibrium constants. In a solution of a silver salt saturated with ammonia, for example, nearly all the silver ion is present as the complex $[Ag(NH_3)_2]^+$, and addition of a chloride-containing solution produces no precipitate; addition of an iodide-containing solution, however, results in the precipitation of silver iodide. Silver iodide is much less soluble than silver chloride, the values for their solubility products being 10^{-16} and 10^{-10} respectively, so what these experiments indicate is that the overall equilibrium constant for the reaction

$$Ag^+ + 2NH_3 \rightleftharpoons [Ag(NH_3)_2]^+$$

is large enough for silver chloride to be soluble in a saturated solution of ammonia whilst silver iodide is unaffected.

Physical methods (such as investigations of colligative properties, electronic or vibrational spectra, solubility, conductivity, or electrode potentials) provide more reliable evidence as well as, in favourable circumstances, leading to values of equilibrium constants for complex formation.

The examples of complexes that are encountered early in the study of chemistry usually contain only one metal ion, which is combined with one or more anionic or neutral *ligands*. Complexes containing anionic ligands include the hexacyanoferrate(II) and hexacyanoferrate(III) ions, $[Fe(CN)_6]^{4-}$ and $[Fe(CN)_6]^{3-}$ respectively, and the neutral bright red nickel(II) dimethylglyoximate, $[Ni(CH_3C(=NO)C(=NOH)CH_3)_2]$, the formation of which is weakly alkaline solution is a sensitive test fcr nickel. The complexes $[Ag(NH_3)_2]^+$ and $[Co(H_2NCH_2CH_2NH_2)_3]^{3+}$ contain uncharged ligands. Complexes containing more than one cation are by no means uncommon, however: partially hydrolysed magnesium salts, for example, contain ions such as $[Mg_2(OH)_3]^+$, and when silver iodide is dissolved in a saturated solution of silver nitrate the ions Ag_2I^+ and Ag_3I^{2+} are produced.

Neutral complexes are usually only sparingly soluble in water but are often readily soluble in organic solvents, e.g. iron(III) acetylacetonate, $[Fe(CH_3COCHCOCH_3)_3]$ (or, as it is often written, $[Fe(acac)_3]$), is a red solid (m.p. 179° C) which is readily extracted from aqueous media by benzene or chloroform. The anion in this substance is formed by the deprotanation of the weak acid acetylacetone, and the formation of the compound in aqueous solution therefore involves both of the equilibria

$$CH_3COCH_2COCH_3 \rightleftharpoons CH_3COCHCOCH_3^- + H^+ \quad K = 10^{-10}$$

and

$$Fe^{3+} + 3CH_3COCHCOCH_3^- \rightleftharpoons [Fe(CH_3COCHCOCH_3)_3] \quad K = 10^{26}$$

The quantity of complex formed therefore depends upon the pH of the solution: if it is too low, H^+ ions compete with Fe^{3+} ions for the anion; if it is too high, the iron is precipitated as hydroxide, for which K_s is about 10^{-37}. There is thus an optimum pH for the extraction of iron(III) from aqueous media by acetylacetone and chloroform. Since most ligands are bases in the Brønsted sense, accurate pH control is of great importance in studies of complex formation.

Formation Constants of Complexes

Let us consider a metal ion M and a ligand L (charges being omitted for the sake of simplicity) forming a series of complexes ML, ML_2, ... ML_n in aqueous solution. Although each stage in complex formation really represents successive replacement of water from the hydration shell of the metal ion, on the convention that the activity of water in such a solution is unity we can disregard it as a reactant or product. For the formation of successive complexes we then have, provided equilibrium is attained in each case,

$$M + L \rightleftharpoons ML \qquad K_1 = \frac{[ML]}{[M][L]}$$

$$ML + L \rightleftharpoons ML_2 \qquad K_2 = \frac{[ML_2]}{[ML][L]}$$

$$ML_{n-1} + L \rightleftharpoons ML_n \qquad K_n = \frac{[ML_n]}{[ML_{n-1}][L]}$$

The equilibrium constants K_1, K_2... K_n are called stepwise formation (or stability) constants. Alternatively we may consider a series of overall equilibrium constants, for which the general symbol β is used. Thus

$$M + L \rightleftharpoons ML \qquad \beta_1 = \frac{[ML]}{[M][L]}$$

$$M + 2L \rightleftharpoons ML_2 \qquad \beta_2 = \frac{[ML_2]}{[M][L]^2}$$

$$M + nL \rightleftharpoons ML_n \qquad \beta_n = \frac{[ML_n]}{[M][L]^n}$$

Obviously

$$\beta_n = K_1 K_2 \ldots K_n$$

For the sake of consistency, it may be noted, we should regard the equilibrium between an acid HL and its ionisation products as the complexing of the proton by the anion

$$H^+ + L^- \rightleftharpoons HL$$

but this representation is seldom employed, most chemists preferring to start from the acid and consider its dissociation.

If it is shown that only one complex, of known formula, is present in a solution containing known total activities (or, approximately, concentrations) of M and L, the formation constant of this complex may be obtained from a determination of the concentration of uncomplexed M or L. This may be made by polarographic or emf measurements if a suitable reversible electrode exists, by pH measurements if the ligand is the anion of a weak acid, or by ion-exchange, spectrophotometric, a nd distribution methods. In general, however, more than one complex is present in a given solution, and then measurements over a wide range of relative concentrations, followed by computational treatment of the experimental data, are necessary for the evaluation of all the consecutive formation constants; details of these operations are given in some of the references listed at the end of the chapter. From measured equilibrium constants, standard free energies of complex formation may be calculated. By making measurements at different temperatures, enthalpies and entropies of complex formation may also be obtained, though since enthalpies of complexing in aqueous solution are usually small and are not quite independent of temperature, it is preferable to obtain them by direct calorimetry.

It is generally found that stepwise formation constants show a progressive decrease as n increases; for the Ni^{2+}–NH_3 system in 2 molar NH_4NO_3 solution at 30° C, for example, a pH study using a glass electrode gave $K_1 = 10^{2.79}$, $K_2 = 10^{2.26}$, $K_3 = 10^{1.69}$, $K_4 = 10^{1.25}$, $K_5 = 10^{0.74}$, $K_6 = 10^{0.03}$, leading to an overall formation constant of the $[Ni(NH_3)_6]^{2+}$ ion $\beta_6 = 10^{8.74}$. Thus the principal complex present in an ammoniacal solution of a nickel(II) salt depends upon the $[NH_3]:[Ni^{2+}]$ ratio; only at high concentrations of ammonia is the pentammine the main species present, and even in the presence of very high concentrations of the ligand, formation of the hexammine complex is incomplete. For the Ni^{2+}–NH_3 system, the enthalpies of successive stages in complex formation are all about –17 kJ mol^{-1} and the relative values of successive formation constants agree very roughly with what might be expected on the basis of a statistical replacement of water by ammonia with the coordination number of the metal ion constant. In other cases, however, an abrupt variation in value of K occurs. Thus in the Ag^+–NH_3 system $K_2 > K_1$, suggesting that coordination of one molecule of ammonia converts the hydration shell round the cation into a monosubstituted tetrahedron or octahedron, but that formation of the second complex produces a total change of structure with formation of the linear $[Ag(NH_3)_2]^+_{(aq)}$ ion.

Factors Affecting the Stabilities of Complexes

We will discuss complexes containing monodentate ligands **only.** The following generalisation exists.

The stabilities of complexes of non-transition metal ions of a given charge normally decrease with increasing size of the cation (e.g. among the alkaline earth metal ions the order of stability is $Ca^{2+} > Sr^{2+} > Ba^{2+}$). Analogous behaviour is found for the series of lanthanide tripositive ions; for the transition metal ions, however, an additional effect (crystal or ligand field stabilisation) is usually involved.

For a given ionic size, increase in charge almost invariably results in a substantial increase in complex stability (e.g. along the series $Li^+ < Mg^{2+} < Al^{3+}$).

If a metal exists in two different oxidation states, the more highly charged ion is the smaller, and the two effects reinforce one another; metals in their higher oxidation states nearly always form more stable complexes than in the lower oxidation states. Occasional exceptions to this rule do occur, for example in the 1,10-phenanthroline and 2,2′-bipyridyl complexes of Fe(II) and Fe(III); but where this is so there are strong reasons for believing that the nature of the ligand (in particular the electron density on it) is not quite the same in complexes of the metal in different oxidation states. It is noteworthy that all ligands that are prone to form stable complexes with metals in low oxidation states (carbon monoxide and cyanide ion, for example, in addition to the two heterocyclic species mentioned already) have relatively low-energy antibonding orbitals available to accept electrons back from the metal ion; they thus differ essentially from the simple inorganic ligands like H_2O, NH_3 and the halide ions, with which we have so far been mainly concerned.

Where their acceptor behaviour towards different ligands is concerned, cations fall into two classes, though the distinction between them is not always clear-cut. For halide ions as ligands, it is found that for the lighter *s* and *p* block cations, metals at the beginnings of the transition series, lanthanides and actinides the usual order of complex stability is fluoride > chloride > bromide > iodide. For tellurium, polonium, the platinum metals, copper, silver, gold, cadmium, mercury and thallium this order is reversed. On the basis of mainly qualitative evidence, cations of the first group also appear to form more stable complexes with amines than with phosphines, and with ethers than with sulphides; these orders are reversed for cations of the second group.

Cations of the first and second groups were called class (*a*) and class (*b*) acceptors by Ahrland, Chatt and Davies, and these terms are widely used. In the wider context of general electron donor (*Lewis base*)–electron acceptor (*Lewis acid*) interaction, class (*a*) cations have been termed 'hard' acids, and class (*b*) cations 'soft' acids by Pearson, who has also classified ligands as 'hard' or 'soft' bases according to whether they are small and their valence electrons are tightly held (e.g. F^-, H_2O, NH_3) or are large and easily polarised (e.g. I^-, CN^-, $(CH_3)_2S$, $(CH_3)_3P$). 'Hard' acids bind more strongly to 'hard' bases, 'soft' acids more strongly to 'soft' bases—the so-called principle of hard and soft acids and bases.

REDOX POTENTIALS

In this section we will discuss the equilibria involving a change in oxidation state in aqueous media, i.e. with transfer of electrons. An example of such a process is the reduction of zinc ions (1*M*) in aqueous solution to metallic zinc:

$$\underset{1M}{Zn^{2+}} + 2e \rightleftharpoons Zn$$

The standard free energy change, ΔG°, associated with this half-reaction is related to the standard redox potential, E°, of the Zn^{2+} (M = 1)/Zn electrode by the equation

$$\Delta G^\circ = -nE^\circ F$$

where n is the number of electrons transferred and F is Faraday's constant. ΔG° and E° are relative to the coıresponding quantities for the half-reaction

$$\underset{1M}{H^+} + e \rightleftharpoons \underset{(1\ \text{atm})}{\tfrac{1}{2}H_2}$$

and the standard hydrogen electrode respectively; E° is found to be −0.76 V. For the cell conventionally written

$$Zn/Zn^{2+}\ (1M)\ ||\ H^+\ (1M)/H_2\ (1\ \text{atm}),\ Pt$$

the chemical reaction is

$$Zn + 2H^+ \rightarrow Zn^{2+} + H_2$$

the emf of this cell is +0.76 V, the standard free energy change is

$$\Delta G^\circ = -2E^\circ F$$

and the equilibrium constant is given by

$$\log K = \frac{2E^\circ F}{2.303RT} = 26$$

A number of standard redox potentials are given in Table 1.5. The more positive the value of E°, the better is the system (with the species on both sides of the half-reaction at unit activity) as an oxidising agent relative to the $H^+/\tfrac{1}{2}H_2$ system.

The way of writing half-reactions used here is now becoming generally accepted, and has the advantage that it gives the correct sign of the potential of the actual electrode relative to that of the standard hydrogen electrode. If the half-reaction is written as an oxidation reaction, e.g.

$$Zn \rightleftharpoons \underset{1M}{Zn^{2+}} + 2e$$

the sign of the standard free energy change is then *necessarily* the opposite of that of the half-reaction

Table 1.5 Some Standard Redox Potentials in Aqueous Solution at 25°C

Electrode reaction	E° (V)
$\frac{1}{2}F_2 + e \rightleftharpoons F^-$	+ 2.9
$\frac{1}{2}S_2O_8^{2-} + e \rightleftharpoons SO_4^{2-}$	+ 2.01
$Ag^{2+} + e \rightleftharpoons Ag^+$	+ 1.98
$Co^{3+} + e \rightleftharpoons Co^{2+}$	+ 1.95
$MnO_4^- + 8H^+ + 5e \rightleftharpoons Mn^{2+} + 4H_2O$	+ 1.51
$\frac{1}{2}Cl_2 + e \rightleftharpoons Cl^-$	+ 1.36
$\frac{1}{2}Cr_2O_7^{2-} + 7H^+ + 3e \rightleftharpoons Cr^{3+} + \frac{7}{2}H_2O$	+ 1.33
$\frac{1}{2}O_2 + 2H^+ + 2e \rightleftharpoons H_2O$	+ 1.23
$[Fe(phen)_3]^{3+} + e \rightleftharpoons [Fe(phen)_3]^{2+}$	+ 1.12
$\frac{1}{2}Br_2 + e \rightleftharpoons Br^-$	+ 1.07
$Ag^+ + e \rightleftharpoons Ag$	+ 0.80
$Fe^{3+} + e \rightleftharpoons Fe^{2+}$	+ 0.77
$\frac{1}{2}I_2 + e \rightleftharpoons I^-$	+ 0.54
$[Fe(CN)_6]^{3-} + e \rightleftharpoons [Fe(CN)_6]^{4-}$	+ 0.36
$Cu^{2+} + 2e \rightleftharpoons Cu^+$	+ 0.34
$AgCl + e \rightleftharpoons Ag + Cl^-$	+ 0.22
$Cu^{2+} + e \rightleftharpoons Cu^+$	+ 0.15
$H^+ + e \rightleftharpoons \frac{1}{2}H_2$	0
$Cr^{3+} + e \rightleftharpoons Cr^{2+}$	− 0.41
$Fe^{2+} + 2e \rightleftharpoons Fe$	− 0.44
$Zn^{2+} + 2e \rightleftharpoons Zn$	− 0.76
$Mn^{2+} + 2e \rightleftharpoons Mn$	− 1.18
$Al^{3+} + 3e \rightleftharpoons Al$	− 1.66
$Mg^{2+} + 2e \rightleftharpoons Mg$	− 2.37
$Na^+ + e \rightleftharpoons Na$	− 2.71
$Ca^{2+} + 2e \rightleftharpoons Ca$	− 2.87
$Li^+ + e \rightleftharpoons Li$	− 3.04

$$\underset{1M}{Zn^{2+}} + 2e \rightleftharpoons Zn$$

and hence the sign of E° must change too. Many chemists find that if they are in any doubt as to whether the sign of a E° value is correct, it is best to consider ΔG°, a negative value for which means that the half-reaction under consideration occurs spontaneously with respect to the half-reaction

$$\underset{1M}{H^+} + e \rightleftharpoons \underset{(1\ \text{atm})}{\tfrac{1}{2}H_2}$$

Any half-reaction may be represented in the form

$$aA + bB + ... + ne \rightleftharpoons pP + qQ + ...$$

The value of the standard redox potential E° then refers to the condition that *all* substances present are at unit activity i.e., $1M$; under non-standard conditions E relative to the standard hydrogen electrode is given by

$$E = E^\circ - \frac{RT}{nF} \ln \frac{[P]^p [Q]^q}{[A]^a [B]^b} ...$$

$$= E^\circ - \frac{0.059}{n} \ln \frac{[P]^p [Q]^q}{[A]^a [B]^b} \text{... at } 25^\circ C$$

(E will be less positive, i.e. ΔG will be less negative and the reduction will be less likely to go spontaneously, in the presence of high concentrations of products.) Thus for the reduction of Zn^{2+}, since the concentration of solid Zn is by convention taken as unity,

$$E = E^\circ - \frac{0.059}{2} \log \frac{1}{[Zn^{2+}]}$$

For the reduction of MnO_4^- to Mn^{2+} in acidic solution, remembering that the concentration of water is taken as unity,

$$E = E^\circ - \frac{0.059}{5} \log \frac{[Mn^{2+}]}{[MnO_4^-][H^+]^8}$$

Thus MnO_4^- being reduced to Mn^{2+} becomes a more powerful oxidant as the hydrogen ion activity increases: MnO_4^- will not oxidise chloride in neutral solution, for example, but it liberates chlorine from concentrated hydrochloric acid.

The potentials for the reduction of water ($[H^+] = 10^{-7}$) to hydrogen and for the reduction of oxygen to water (the reverse of the oxidation of water to oxygen) are of particular importance in aqueous solution chemistry, since they provide general guidance concerning the nature of chemical species that can exist under aqueous conditions. For the reduction

$$H^+ + e \rightleftharpoons \tfrac{1}{2}H_2$$

$$E = E^\circ - 0.059 \log \frac{p_{H^1}^{\frac{1}{2}}}{[H^+]}$$

since hydrogen is almost an ideal gas. For neutral water in equilibrium with hydrogen gas at atmospheric pressure, $E = -0.41$ V; at $[H^+] = 10^{-14}$, $E = -0.83$ V. Thus any couple with E° more negative than -0.41 V should reduce water under a pressure of one atmosphere of hydrogen, and any couple with E° more negative than -0.83 V should reduce molar alkali under the same conditions. The potential of -0.83 V for the $H^+/\tfrac{1}{2}H_2$ electrode in molar alkali is of only limited importance in isolation; many metal ion/metal systems that should reduce water under these conditions are prevented from doing so by formation of a coating of hydroxide or hydrated oxide; others which are less powerfully reducing (e.g. Zn^{2+}/Zn) bring about reduction because they are modified by complexing (e.g. conversion of the Zn^{2+} ions into a stable hydroxo-complex $[Zn(OH)_4]^{2-}$).

For the reduction of oxygen the relevant reaction and standard potential at $[H^+] = 1$ are

$$\tfrac{1}{2}O_2 + 2H^+ + 2e \rightleftharpoons H_2O \qquad E = +1.23 \text{ V}$$

Then under non-standard conditions

$$E = E^\circ - \frac{0.059}{2} \log \frac{1}{p_{O_2}^{\frac{1}{2}}[H^+]^2}$$

For $p_{O_2} = 1$, $E = 1.23 - 0.059$ pH; for neutral water under atmospheric pressure of oxygen E is thus $+0.83$ V, and for molar alkali under the same conditions, $+0.41$ V. So from the standpoint of thermodynamics, oxygen in the presence of water should oxidise any system with E° less positive than $+1.23$ V at pH 0, $+0.83$ V at pH 7, and + 0.42 V at pH 14. Conversely, any system with E° more positive than 1.23 V should oxidise water at pH 0, and so on. For both the reduction and the oxidation of water, the other system refers, of course, to one in which all species involved in the appropriate half-reaction are present at unit activity (otherwise we would not be considering E° values). In reality, of course, most of experimental chemistry involving changes in oxidation states is carried out by taking an almost pure reductant and allowing it to react with an almost pure oxidant. Thus although a solution combining Cr^{3+} and Cr^{2+} both at unit activity (E° Cr^{3+}/Cr^{2+} = – 0.41 V) in water under one atmosphere pressure of hydrogen is in equilibrium, addition of a chromium(II) salt to water should lead to liberation of hydrogen.

THE STABILISATION OF OXIDATION STATES

For the half-reaction

$$Ag^+ \ (1M) + e \rightleftharpoons Ag$$

E° is +0.80 V. If the concentration of Ag^+ ions is reduced, E° will become less positive and reduction to the metal will be less easy. We may then say that the oxidation state Ag(I) has been stabilised. This may be brought about by removal of the ion either by formation of a stable complex or by precipitation as a sparingly soluble sat. If K is the overall equilibrium constant for the process which reduces the silver ion concentration, we can calculate the E° value for the new half-reaction from the relationships

$$-\Delta G^\circ = RT \ln K = n(\Delta E^\circ)F$$

Silver chloride, for example, has a solubility product of 10^{-10}, from which we can easily show that for the half-reaction

$$AgCl_{(s)} + e \rightleftharpoons Ag + \underset{1M}{Cl^-}$$

E° is +0.22 V: it is harder to reduce silver chloride than the hydrated silver ion. Silver iodide is less soluble than silver chloride, but it is much more soluble in aqueous potassium iodide than silver chloride is in aqueous potassium chloride; thus the iodo complexes of silver are much more stable than the chloro complexes. For the complex AgI_3^{2-} the overall formation constant is found to be 10^{14}. Combination of this value with E° for the Ag^+/Ag electrode gives

$$AgI_3^{2-} + e \rightleftharpoons Ag + 3I^- \qquad E^\circ = -0.02 \text{ V}$$

Again silver (I) has been stabilised against reduction, but this time to a greater degree: silver in the presence of AgI_3^{2-} and I^- both at unit activity is as powerful a reductant as hydrogen in the presence of hydrogen ion at unit activity. If we heat powdered silver with a concentrated solution of hydrogen iodide, evolution of hydrogen occurs.

The modification of the relative stabilities of two oxidation states of a metal, both subject to removal by precipitation or complexing, may be treated in a similar way. For example, manganese(III) is a powerful oxidant:

$$Mn^{3+} + e \rightleftharpoons Mn^{2+} \qquad E^\circ = +1.51 \text{ V}$$

In alkaline medium both cations are precipitated, but Mn^{3+} much more completely than Mn^{2+}, the solubility products of $Mn(OH)_3$ and $Mn(OH)_2$ being 10^{-36} and 10^{-13} respectively, whence at $[OH^-] = 1$

$$Mn(OH)_3 + e \rightleftharpoons Mn(OH)_2 + OH^- \qquad E = +0.15 \text{ V}$$

Thus although oxygen at $[H^+] = 1$ cannot oxidise Mn^{2+}, oxygen at $[OH^-] = 1$, for which $E = +0.41$ V (as we have already seen) can and does oxidise $Mn(OH)_2$. Most transition metals resemble manganese in being more stable in higher oxidation states in alkali than in acid because the hydroxide of the metal in its higher oxidation state is much less soluble than the hydroxide of the metal in its lower oxidation state.

Analogous principles apply when two oxidation states of a metal are both stabilised by complexing: it is generally found that the higher oxidation state is stabilised to a greater degree. Thus comparison of

$$Co^{3+} + e \rightleftharpoons Co^{2+} \qquad E^\circ = +1.95\text{V}$$

and

$$[Co(NH_3)_6]^{3+} + e \rightleftharpoons [Co(NH_3)_6]^{2+} \quad E^\circ = +0.10 \text{ V}$$

shows that the overall formation constant of the $[Co(NH_3)_6]^{3+}$ ion must be about 10^{31} times as great as that of the $[Co(NH_3)_6]^{2+}$ ion. A similar comparison of

$$Fe^{3+} + 3 \rightleftharpoons Fe^{2+} \qquad E^\circ = +0.77 \text{ V}$$

and

$$[Fe(CN_6)_6]^{3-} + e \rightleftharpoons [Fe(CN)_6]^{4-} \qquad E^\circ = +0.36 \text{ V}$$

leads to the conclusion that the overall formation constant of the $[Fe(CN)_6]^{3-}$ ion is 10^7 times that of the $[Fe(CN)_6]^{4-}$ ion.

There are some reactions in which an element in one oxidation state is converted into two other oxidation states: examples of such *disproportionations* are:

$$2Cu^+ \rightleftharpoons Cu + Cu^{2+}$$

and

$$3MnO_4^{2-} + 4H^+ \rightleftharpoons 2MnO_4^- + MnO_2 + 2H_2O$$

The former of these may be seen to take place when copper (I) sulphate (prepared by interaction of copper (I) oxide and dimethyl sulphate) is added to water; the latter occurs when acid is added to a solution of potassium manganate (VI). Equilibrium constants for such disproportionations may be calculated from potential data. The reaction

$$2Cu^+ \rightleftharpoons Cu + Cu^{2+}$$

for example, for which $K = 10^6$, is the sum of

$$Cu^+ \rightleftharpoons Cu^{2+} + e$$

and

$$Cu^+ + e \rightleftharpoons Cu$$

Species which are unstable with respect to disproportionation, such as Cu^+ in aqueous solution, may, like other oxidation states, be stabilised under suitable conditions—in this case by precipitation as a sparingly soluble species (e.g. as CuCl) or by complexing (e.g. as $[Cu(CN)_4]^{3-}$). In the case of MnO_4^{2-} all that is necessary is to remove the H^+ ions involved in bringing about disproportionation by keeping the solution alkaline. But even this may be considered as yet another example of complexing: neutralisation in aqueous solution is only hydroxide ion complexing of the proton, for which, on the convention that the activity of the water is unity, $K = 10^{14}$.

E° VALUE AND OXIDATION STATE

For an element which exists in several different oxidation states in aqueous solution, many half-reactions need to be tabulated to give a clear picture of its solution chemistry. Manganese, for example, exhibits well defined oxidation states of II, III, IV, VI and VII in aqueous media. The following standard potentials may be determined experimentally:

$$Mn^{2+} + 2e \rightleftharpoons Mn \qquad E^\circ = -\ 1.18\ V$$

$$Mn^{3+} + e \rightleftharpoons Mn^{2+} \qquad E^\circ = -\ 1.51V$$

$$MnO_2 + 4H^+ + 2e \rightleftharpoons Mn^{2+} + 2H_2O \qquad E^\circ = +\ 1.23\ V$$

$$MnO_4^- + 8H^+ + 5e \rightleftharpoons Mn^{2+} + 4H_2O \qquad E^\circ = +\ 1.51\ V$$

$$MnO_4^- + e \rightleftharpoons MnO_4^{2-} \qquad E^\circ = +\ 0.56\ V$$

These potentials may be used to derive E° for other half-reaction such as

$$MnO_4^- + 4H^+ + 3e \rightleftharpoons MnO_2 + 2H_2O \qquad E^\circ = +\ 1.69\ V$$

In seeking to obtain an overall view of the relative stabilities of the different oxidation states, it is useful to present all these data on a diagram whose main features are apparent at a glance, and to this end the so-called potential diagrams and oxidation state diagrams have been devised.

The potential diagram of manganese at $[H^+] = 1$ is shown in Fig. 1.1.

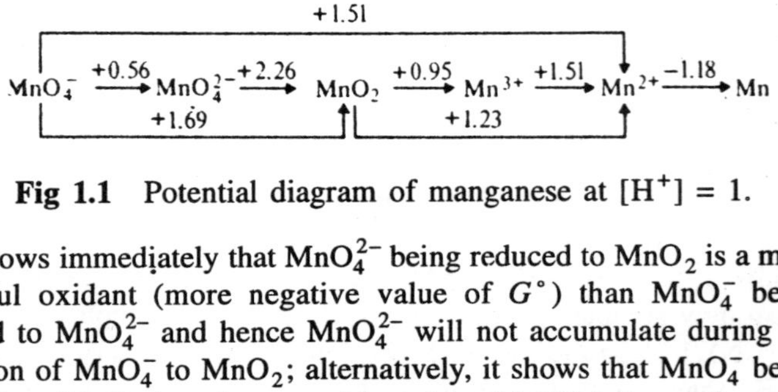

$$MnO_4^- \xrightarrow{+0.56} MnO_4^{2-} \xrightarrow{+2.26} MnO_2 \xrightarrow{+0.95} Mn^{3+} \xrightarrow{+1.51} Mn^{2+} \xrightarrow{-1.18} Mn$$

$MnO_4^- \rightarrow Mn^{2+}$: +1.51; $MnO_4^- \rightarrow MnO_2$: +1.69; $MnO_2 \rightarrow Mn^{2+}$: +1.23

Fig 1.1 Potential diagram of manganese at $[H^+] = 1$.

It shows immediately that MnO_4^{2-} being reduced to MnO_2 is a more powerful oxidant (more negative value of G°) than MnO_4^- being reduced to MnO_4^{2-} and hence MnO_4^{2-} will not accumulate during the reduction of MnO_4^- to MnO_2; alternatively, it shows that MnO_4^- being reduced to MnO_2 is a better oxidant than MnO_4^- being reduced to MnO_4^{2-}; or we can say that MnO_4^{2-} at pH 0 is unstable with respect to disproportionation into MnO_4^- and MnO_2. Similarly Mn^{3+} is unstable with respect to disproportionation into MnO_2 and Mn^{2+} at pH 0. It is important to note that protons are involved in some stages, they are, however, not explicitly shown. Nevertheless they do enter into many redox equilibria. At pH values other than zero, those potentials which relate to half-reactions involving protons will change to extents depending on the number of protons involved, whilst the others will remains constant. Thus for an element forming insoluble oxides and oxo-anions, a new potential diagram is needed for every change in hydrogen ion concentration. The incorporation of data for systems involving complexes or insoluble salts is usually not attempted.

In the common graphical method of summarising redox relationships the standard free energy change ΔG° for the formation of M(Z) from M(0), where Z is the oxidation state, is plotted against increasing Z. This form of plot is often used even where half-reactions are written,

$$\text{oxidised form} + \text{ne} \rightleftharpoons \text{reduced form}$$

In the oxidation state diagram for manganese at $[H^+] = 1$, the lowest point on it (Mn(II)) represents the most stable oxidation state. Those half-reactions which have negative ΔG° are represented by downward slopes, those with positive ΔG° by upward slopes; if $\Delta G^\circ / F$ has been plotted against Z, the slope gives the value of E°. Any state represented by a 'convex' point is thermodynamically unstable with respect to the states on either side of it; the reverse is true of a state represented by a 'concave' point. It must again be emphasised that, since protons are involved in many of the changes represented, the diagram would look quite different at other hydrogen ion concentrations.

2

Acids and Bases

INTRODUCTION

Criteria for characterizing acids and bases used in the beginning were mainly experimental. Thus, Boyle, in the sixteenth century, defined acids as substances whose aqueous solution turned blue litmus red, neutralised bases, reacted with active metals evolving hydrogen, and tasted sour. Bases were similarly defined, in 1774, by Rouelle as substance whose aqueous solution turned red litmus blue, neutralised acids, tasted bitter, and gave soapy touch. The first departure from a strictly experimental definition of an acid came with the advent of Lavoisier's oxygen concept. Lavoisier (1787) proposed that the peculiar properties of an acid could be attributed to the presence of oxygen, as the acids H_2SO_4, H_3PO_4 have oxygen. The oxygen concept was abandoned when Davy (1811) showed that it was not necessary for an acid to contain oxygen (HCl, H_2S), and further, many binary oxygen compounds (CaO, K_2O) did not possess acidic properties. So Sir Davy, finally in 1816, expressed that hydrogen is peculiar in all acidic species. Liebig (1838), the chief protagonist of this new concept, defined acids as substances possessing one or more hydrogen atom replaceable by metals and thus the hydrogen concept become popular.

THE ARRHENIUS CONCEPT OF ACIDS AND BASES

According to Arrhenius (1884), an acid is any hydrogen-containing compound which yields hydrogen ions (H^+) in aqueous solution and a base is any hydrogen-containing compound which gives hydroxyl ions (OH^-) in aqueous solution. Thus,

$$\underset{\text{acid}}{HCl(aq)} \rightleftharpoons H^+(aq) + Cl^-(aq)$$

$$\underset{\text{base}}{NaOH(aq)} \rightleftharpoons Na^+(aq) + OH^-(aq)$$

The process of neutralization of an acid by a base can be represented as

$$H^+ + OH^- \rightleftharpoons H_2O$$

This concept explained the constant heat of neutralisation of any strong acid with any strong base as all these reactions involve only the combination of hydrogen and hydroxyl ions.

One of the most significant achievements of Arrhenius concept was its ability to explain the catalytic properties of acids. Arrhenius theory of electrolytic dissociation correlated the catalytic actions of acids with the concentration of hydrogen ion.

Limitations

In spite of the successes of Arrhenius definition, several shortcomings of this definition soon became apparent. The definition of acids and bases, for example, in terms of their behaviour in aqueous medium, limited the scope of acid-base concepts.

1. Towards the end of the nineteenth century, it was shown that acid-base reactions could be carried out in non-aqueous media, such as liquid ammonia and liquid sulphur dioxide, just as medium. Arrhenius theory, however, denied the occurrence of such reactions in non-aqueous solvents.
2. Further, these concepts were not applicable to reactions in the gas phase where no solvent is present. For example, HCl which is acidic in aqueous phase, remains unexplained when exists in the gaseous form.
3. The restriction of the term base to only hydroxyl compounds was another serious limitation of the Arrhenius definition.
4. Many substances such as ammonia and pyridine were known to exhibit basic properties although they contain no hydroxyl group.
5. The neutralisation process is limited to those reactions which can occur in aqueous solutions only, although reactions involving salt formation do occur in many other solvents and even in the absence of solvents.

BRONSTED–LOWRY CONCEPT

Of the several concepts proposed to explain the properties and reactions of acids and bases, the one originated independently in 1923 by Johannes Nicolaus Brønsted of Denmark and Thomas Martin Lowry of England is possibly the most useful. According to this approach, which, unlike the older Arrhenius approach, is not limited to aqueous solutions but is applicable to all proton (H^+)-containing systems, and acid is a proton donor, and a base is a proton acceptor; the strength of an acid or base is correlated with its tendency to donate or accept protons, respectively. Acids may be positive ions (cations), neutral molecules, or anions, while bases may be negative ions (anions) as well as neutral molecules.

Bronsted-Lowry acids and bases are of the following types :

Type	Acid	Base
Molecular	HCl, HBr, $HClO_4$, H_2SO_4, H_3PO_4, H_2O	NH_3, N_2H_4, amines, H_2O
Cationic	NH_4^+, $[Fe(H_2O)_6]^{3+}$ $[Al(H_2O)_6]^{3+}$	$[Fe(H_2O)_5(OH)]^{2+}$ $[Al(H_2O)_5(OH)]^{2+}$
Anionic	HS^-, HCO_3^-, HSO_4^-, $H_2PO_4^-$	Cl^-, Br^-, OH^-, HSO_4^-, CO_3^{2-}, SO_4^{2-}

Conjugate acid-base pairs

In a typical acid-base reaction

$$HA + B \rightleftharpoons A^- + BH^+$$

HA being a proton donor is an acid and B being a proton acceptor is a base. In the reaction HB^+ being a proton donor is an acid and A^- being a proton acceptor is a base. In order to distinguish between the two acids and two bases, those on the right-hand side of the equation are referred to as the conjugate acid and the conjugate base. Thus BH^+ is a conjugate acid of the bases B and A^- is a conjugate base of acid HA. HA–A^- and HB^+–B constitute two conjugate acid-base pairs. The acid and base in each conjugate differ by a proton. For an acid to exhibit acidic properties there must be a proton acceptor (base) present. Some examples are:

Acid$_1$		Base$_2$		Conjugate acid$_2$		Conjugate base$_1$
HCl	+	H_2O	$\rightleftharpoons$	H_3O^+	+	Cl^-
HSO_4^-	+	NH_3	$\rightleftharpoons$	NH_4^+	+	SO_4^{2-}
NH_4^+	+	H_2O	$\rightleftharpoons$	H_3O^+	+	NH_3

Thus, according to the Bronsted-Lowry concept neutralization is a process of transfer of proton from an acid to a base. For example :

$$CH_3COOH + NH_3 \rightleftharpoons NH_4^+ + CH_3COO^-$$

$$NH_4^+ + S^{2-} \rightleftharpoons HS^- + NH_3$$

$$[Fe(H_2O)_6]^{3+} + H_2O \rightleftharpoons H_3O^+ + [Fe(H_2O)_5(OH)]^{2+}$$

Basicity of an Acid is defined as the number of H^+ ions furnished by a molecule of an acid. An acid may be classified according to its basicity; i.e. acids are monobasic acids such as, HCl, HNO_3, etc. dibasic acids such as, H_2SO_4, H_2CO_3, etc. and tribasic or triprotic acids, like H_3PO_4, as they liberate one, two and three H^+ ions respectively.

Acidity of a base may be defined as the number of OH^- ions furnished by a molecule of a base. A base can be monoacid like $NaOH$, NH_4OH, etc., diacid or dihydroxic like $Ba(OH)_2$, $Mg(OH)_2$, etc. and triacid or trihydroxic, $Fe(OH)_3$, $Al(OH)_3$, etc. when it contains one, two and three hydroxide ions respectively.

Water can act both as an acid as well as base. It acts as a base (proton acceptor) towards HCl and as an acid (proton donor) toward NH_3.

HCl	+	H_2O	$\rightleftharpoons$	H_3O^+	+	Cl^-
Acid		Base		Acid		Base
NH_3	+	H_2O	$\rightleftharpoons$	NH_4^+	+	OH^-
Base		Acid		Acid		Base

An acid-base reaction always proceeds in the direction of formation of the weaker acid and the weaker base.

Consider the equilibrium reaction :

$$HA + H_2O \rightleftharpoons H_3O^+ + A^-$$

where HA is a strong acid, the fact that the equilibrium shifts to the right implies that A^- has very little affinity for protons. Therefore, A^- must be a weak base. Thus we may infer that the conjugate base of a strong acid is always a weak base and the conjugate base of a weak acid is always a strong base.

A table of Bronsted acids and their conjugate bases (the acids from which one proton has been removed) in order of decreasing acid strength is very useful. The acid-base table is similar to the already familiar activity (EMF) series and can be used similarly. Just as the EMF table consists of couples, each consisting of a reduced and oxidized form, this acid-base table consists also of couples, each consisting of an acid and a base (conjugate acid-base pairs). Just as the EMF series is conventionally given in terms of tendency toward electron loss (oxidation), so the acid-base table is conventionally given in terms of tendency toward proton loss (acidity). Just as E° values give the tendency toward electron loss, so K_a values give the tendency toward proton loss. Just as the best reducing agents are found in the upper left-hand portion of EMF table and the best oxidizing agents in the lower right-hand portion, so the strongest acids are found in the upper left-hand portion of the acid-base table and the strongest bases on the lower right-hand portion.

The analogy between the EMF series and the Bronsted-Lowry concept stems from the fact that both use half-reactions to store thermodynamic data, E° or K_a values, respectively. The procedure suggested here is based on the competition for a real or "imagined" intermediate, i.e., electrons or protons, respectively. Thus this approach can similarly be applied to the Lux-Flood acid-base concept or Ellingham diagrams, in which the intermediates are oxide ions or oxygen atoms, respectively. In our general chemistry course we discuss electrochemistry before acids and bases. For courses treating these topics in the opposite order the analogy still holds, but the procedure suggested should be reversed, i.e., the EMF series should be approached as analogous to the already discussed Brønsted-Lowry concept.

If we consider two given couples in the EMF table, the reducing agent of the couple higher in the table will reduce the oxidizing agent of the couple lower in the table, i.e., the "higher" reaction will go to the right, while the "lower" reaction goes to the left. Similarly, if we consider two given conjugate acid-base pairs in the acid-base table, the acid of the pair higher in the table will transfer a proton to the base of the pair lower in the table, i.e., the "higher" reaction goes to the right while the "lower" reaction goes to the left. In other words, in all acid-base reactions, stronger acids react to form weaker acids, while *simultaneously*, stronger bases react to form weaker bases.

A reduction must always accompany an oxidation, since while the reducing agent gives up its electrons (is oxidized), the oxidizing agent accepts these electrons (is reduced). Similarly, in any acid base reaction, the stronger acid gives up its protons, while the stronger base *simultaneously* accepts them (a weaker acid and a weaker base thus being formed).

Note that the table is arranged so that *conjugate acid base pairs are opposite each other.* Furthermore, *coming down the table acids become weaker, but bases become stronger.* Observe also that acid constants (K_a values) become numerically smaller coming down the table.

Because the stronger the Brønsted acid, the weaker the corresponding conjugate base, and vice versa, the strongest acids appear in the upper left-hand portion of the table, and the strongest bases appear in the lower right-hand portion. An acid-base reaction consists of the transfer of a proton from an acid to a base. The acid therefore forms its conjugate base, and the base forms its conjugate acid. A substance reacts as an acid only if a base is present to accept the proton of the acid, while a substance reacts as a base only if an acid is present to donate a proton to the base.

An acid-base reaction is therefore a competition between two bases for a proton. If the stronger (S) of the two acids and the stronger of the two bases are reactants (appear on the left side of the equation), the reaction is said to proceed to a *large* extent:

$$\underset{\substack{\text{hydrogen chloride} \\ SA_1}}{HCl} + \underset{\substack{\text{water} \\ SB_2}}{H_2O} \rightarrow \underset{\substack{\text{hydronium ion} \\ WA_2}}{H_3O^+} + \underset{\substack{\text{chloride ion} \\ WB_1}}{Cl^-} \quad (1)$$

where the numerical subscripts are used to identify the corresponding conjugate acid-base pairs. If the weaker (W) of the two acids and the weaker of the two bases are reactants (appear on the left side of the equation), the reaction is said to proceed to only a *small* extent:

$$\underset{\substack{\text{acetic acid} \\ WA_1}}{HC_2H_3O_2} + \underset{\substack{\text{water} \\ WB_2}}{H_2O} \rightarrow \underset{\substack{\text{hydronium ion} \\ SA_2}}{H_3O^+} + \underset{\substack{\text{acetate ion} \\ SB_1}}{C_2H_3O_2^-} \quad (2)$$

Equations (1) and (2) therefore agree with our common knowledge that in aqueous solution hydrochloric acid is a strong acid (completely dissociated), while acetic acid is a weak acid. (Its aqueous solution consists largely of undissociated molecules.)

Since any acid stronger than hydronium ion will donate its proton to water to form the hydronium ion, these acids appear equally strong in water (the so-called leveling effect), that is, the hydronium ion is the strongest acid that can exist in aqueous solution. In order to differentiate between the strengths of different acids that are stronger than hydronium ion, a solvent that is a stronger acid than water must be used.

In eqs (1) and (2) in which acids react with water, the water molecule behaves as a base. When bases react with water, on the other hand, the water molecule behaves as an acid, as shown in the following two equations (A substance that can function as an acid in one reaction and as a base in a different reaction is said to be amphiprotic) :

O^{2-}	+	H_2O	→	OH^-	+	OH^- (or $2OH^-$)	
oxide ion		water		hydroxide ion		hydroxide ion	(3)
SB_1		SA_2		WA_1		WB_2	

This reaction proceeds to a large extent, and therefore any metal oxide that dissolves in water gives a solution of the metal hydroxide.

NH_3	+	H_2O	→	NH_4^+	+	OH^-	
ammonia		water		ammonium ion		hydroxide ion	(4)
WB_1		WA_2		SA_1		SB_2	

The fact that this reaction proceeds only to a small extent agrees with the well-known fact that ammonia is a weak base. (Its aqueous solution consists largely of undissociated molecules.)

Since any base stronger than hydroxide ion will accept a proton from water to form the hydroxide ion, these bases appear equally strong in water (the leveling effect); i.e., the hydroxide ion is the strongest base that can exist in aqueous solution. In order to differentiate between the strengths of different bases that are stronger than hydroxide ion, a solvent that is a stronger base than water must be used.

Hydrolysis (reaction of an ion of a salt with water to produce either H_3O^+ or OH^- ion) is simply an acid base reaction and is therefore readily accounted for according to the Brønsted-Lowry approach. A salt will hydrolyze to produce an acidic solution if it contains an anion that is a Bronsted base. If the cation is stronger as a Bronsted acid than the anion is as a Brønsted base, the resulting solution will be acidic. If this situation is reversed, the solution will be basic. The following three equations summarize the possible cases:

Acidic Solution; Small Extent

$$\underset{\text{hexaaquaaluminum ion}}{Al(H_2O)_6^{3+}} + \underset{\text{water}}{H_2O} \rightarrow \underset{\text{pentaaquahydroxoaluminum ion}}{AlOH(H_2O)_5^{2+}} + \underset{\underset{SA_2}{\text{hydronium ion}}}{H_3O^+} \quad (5)$$

Basic Solution; Small Extent

$$\underset{\underset{WB_1}{\text{carbonate ion}}}{CO_3^{2-}} + \underset{\underset{WA_2}{\text{water}}}{H_2O} \rightarrow \underset{\underset{SA_1}{\text{hydrogencarbonate ion}}}{HCO_3^-} + \underset{\underset{SB_2}{\text{hydroxide ion}}}{OH^-} \quad (6)$$

Neutral Solution; Small Extent

$$\underset{\underset{WA_1}{\text{ammonium ion}}}{NH_4^+} + \underset{\underset{WB_2}{\text{acetate ion}}}{C_2H_3O_2^-} \rightarrow \underset{\underset{SB_1}{\text{ammonia}}}{NH_3} + \underset{\underset{SA_2}{\text{acetic acid}}}{HC_2H_3O_2} \quad (7)$$

An aqueous solution of $NH_4C_2H_3O_2$ is neutral even though it hydrolyzes because acetic acid is as strong an acid ($K_a = 1.8 \times 10^{-5}$) as ammonia is a base ($K_a = 1.8 \times 10^{-5}$).

Levelling and Differentiating Solvents

All mineral acids donate proton to water and are completely ionized in it. Water is called here a levelling solvent, because it makes all mineral acids of almost the same strength. Several mineral acids are, however, only partially ionized in glacial acetic acid medium because acetic acid is a poor proton acceptor but rather a better proton donor. Acetic acid, therefore, acts as a differentiating solvent towards mineral acids. The relative strengths of some mineral acids in acetic acid medium are:

$$HClO_4 > HBr > H_2SO_4 > HCl > HNO_3$$

It is also evident that a solute which acts as a weak base in water behaves as a strong base in acetic acid medium. Acetic acid thus exerts a levelling effect on bases. Weak bases such as acetamide and acetanilide which cannot be titrated in aqueous medium with acids can, however, be easily titrated in glacial acetic acid medium. Perchloric acid, the most highly dissociated of the strong acids, in acetic acid, is commonly used as titrant for weak bases.

LUX-FLOOD CONCEPT

Lux (1939) observed that the acid-base reactions are also feasible in oxide systems without the aid of protons. This concept was extended by Flood (1947) and applied to non-protonic systems which were not covered by the Bronsted-Lowry concept. In terms of the Lux-Flood definition, a base (CaO, BaO or Na_2O) is an oxide ion donor and an acid (SiO_2, CO_2, or P_4O_{10}) is an oxide ion acceptor as shown in the equation :

$$\underset{\text{base}}{B} \rightleftharpoons \underset{\text{oxide ion}}{O^{2-}} + \underset{\text{acid}}{A}$$

Some acid-base reactions, in terms of the Lux-Flood concept, may be represented as :

$$CaO + SiO_2 \rightarrow CaSiO_3$$
$$BaO + CO_2 \rightarrow BaCO_3$$
$$\underset{\text{base}}{6\,Na_2O} + \underset{\text{acid}}{P_4O_{10}} \rightarrow 4\,Na_3PO_4$$

Substances may be classified as amphoteric in terms of Lux-Flood definition if they show a tendency both to take up and give up oxide ions

$$\underset{\text{base}}{ZnO} + S_2O_7^{2-} \rightarrow Zn^{2+} + 2SO_4^{2-}$$
$$Na_2O + \underset{\text{acid}}{ZnO} \rightarrow 2Na^+ + ZnO_2^{2-}$$

The Lux-Flood concept of acid-base reactions in oxide systems can be extended to any negative ion, i.e., the halides, sulphides. This is illustrated by the following reactions :

$$3NaF + AlF_3 \xrightarrow{\text{high temp.}} 3Na^+ + AlF_6^{2-}$$
$$\underset{\text{base}}{Na_2S} + \underset{\text{acid}}{CS_2} \longrightarrow \underset{\text{acid}}{2Na^+} + \underset{\text{base}}{CS_3^{2-}}$$

The Solvent System Concept of Acids and Bases

The protonic definition of acids and bases can be extended to reaction in non-aqueous solvents containing hydrogen, such as NH_3, N_2H_4, HF, H_2SO_4, CH_3COOH, HCN, etc. This definition, however cannot be applied to non-protonic solvents. Acid-base behaviour can be observed in systems where a proton plays no part in the reactions at all. It would, therefore, be desirable to have definition of acids and bases which is applicable to protonic as well as non-protonic solvents. Franklin developed the solvent system definition of acids and bases in liquid ammonia. This definition can be extended to acid-base behaviour in other protonic solvents.

Cady and Elsey (1928) proposed *that substances which increase the concentration of the cation characteristic of the solvent are acids, whereas substances which increase the concentration of the anion characteristic of the solvent are bases.* Thus, in water, substances providing protons are acids and those providing hydroxyl ions are bases. Similarly, in liquid ammonia, ammonium salts are acids because they provide ammonium ions, NH_4^+, and amides are bases because they provide amide ions, NH_2^-.

$$
\begin{aligned}
2H_2O &\rightleftharpoons H_3O^+ + OH^- \\
2NH_3 &\rightleftharpoons NH_4^+ + NH_2^- \\
2CH_3COOH &\rightleftharpoons CH_3COOH_2^+ + CH_3COO^- \\
2H_2SO_4 &\rightleftharpoons (H_2SO_4H)^+ + HSO_4^- \\
2SO_2 &\rightleftharpoons SO^{2+} + SO_3^{2-} \\
\text{solvent} & \qquad \text{acid} + \text{base}
\end{aligned}
$$

Though the solvent system definition of acids and bases is useful, it has certain limitations. For instance, in a medium of low dielectric constant, the existence of ions can not be conceived. Furthermore the concept lays too much emphasis on ionic mechanism of the reaction.

LEWIS CONCEPT

In 1923, Lewis proposed a new concept of acids and bases based on electronic theory of valency. According to him an acid is any species which can accept an electron pair while a base is any species which can donate an electron pair. Thus, in the Lewis system, an acid is an electron pair acceptor, whereas a base is an electron pair donor. It accounts all reactions involving hydrogen ion, oxide ion, or solvent interactions, formation of acid-base adducts such as K_3NBF_3 and all coordinate components.

The process of neutralisation involves the formation of a new coordinate-covalent bond between the acid and the base. A proton is, thus, a Lewis acid, and ammonia a Lewis base because the former can accept an electron pair from the latter yielding ammonium ion.

$$H^+ + \overset{x}{\underset{x}{}}NH_3 \rightarrow (H \leftarrow NH_3)^+$$

HCl reacts with water and pyridine as shown below :

$$HCl + H_2O \rightarrow (H_3O)^+ + Cl^-$$

$$HCl + C_6H_5N \rightarrow (C_6H_5NH)^+ + Cl^-$$

In the reaction between BF_3, and NH_3, boron accepts the lone pair from nitrogen forming a covalent bond :

$$BF_3 + \overset{x}{\underset{x}{}}NH_3 \rightarrow [F_2B \leftarrow NH_3]$$

Lewis concept does not differ from Bronsted-Lowry concept with respect to substances classified as bases. A Lewis base is capable of donating electron pair to a proton. Being proton acceptor it is also a Bronsted base. Thus both definitions label the following species as bases.

$$:\ddot{\underset{..}{O}}H^-,\ H_2\ddot{O}:,\ :\ddot{\underset{..}{Cl}}:^-,\ :\ddot{\underset{..}{Br}}:^-,\ :NH_3$$

However, many species which are acids in terms of Lewis definition cannot be termed so according to Bronsted definition. Some examples are sulphur trioxide and halides of boron, aluminium, iron (ferric) and zinc. For example in the reaction, BF_3 does not accept H^+ ions but a pair of electrons from NH_3. Thus BF_3 is a Lewis acid and NH_3 a Lewis base.

Classification of Lewis Acids

Lewis acids may be classified as under :

1. Simple cations. All simple cations are Lewis acids but their strength as acids varies. Some reactions of the simple cations are :

$$Cu^{2+} + 4NH_3 \rightarrow [Cu(NH_3)_4]^{2+}$$

$$Co^{2+} + 6H_2O \rightarrow [Co(H_2O)_6]^{2+}$$

$$\left[\begin{array}{ccc} & OH_2 & \\ H_2O & \downarrow & OH_2 \\ & Co & \\ H_2O & \uparrow & OH_2 \\ & OH_2 & \end{array}\right]^{2+}$$

greater the change on the cation or smaller the ionic radius greater will be the Lewis acid strength or coordinating ability of the simple cations. Thus, Lewis acid strength increases as

$$Fe^{3+} > Fe^{2+}$$
$$Mg^{2+} > Na^{+}$$
$$Li^{+} > Na^{+} > K^{+}$$

2. Molecules containing a central atom with an incomplete octet. Lewis acids of this class are the electron-deficient molecules such as the halides of boron and aluminium acids SO_3:

:Ö: :X: :X:
S :Ö: Al :X: B :X:
:Ö: :X: :X:

$$BF_3 + O(C_2H_5)_2 \rightarrow F_3B \leftarrow O(C_2H_5)_2$$
$$AlCl_3 + NC_5H_5 \rightarrow Cl_3Al \leftarrow NC_5H_6$$

3. Molecules containing vacant d-orbitals on the central atom. Some examples of Lewis acids of this type are SiX_4, GeX_4, SnX_4, TiX_4, PF_3, PF_5, $TeCl_4$, and SeF_4. These halides tend to form adducts

$$SnCl_4 + 2Cl^- \rightarrow [SnCl_6]^{2-}$$
$$SbF_3 + 2F^- \rightarrow [SbF_5]^{2-}$$

SiF_4 and GeF_4 form 1:2 adducts with a variety of oxygen and nitrogen donors.

Halides of this type are hydrolysed by water to form an oxyacid of the central element and the appropriate hydrogen halide. The hydrolytic reactions take place presumably through the intermediate formation of unstable adducts with H_2O.

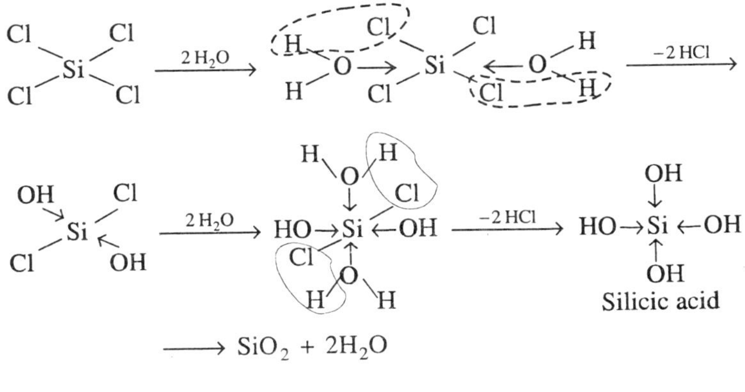

$$SiCl_4 \xrightarrow{2H_2O} H_2O \rightarrow SiCl_4 \leftarrow OH_2 \xrightarrow{-2HCl} Si(OH)_2Cl_2$$

$$Si(OH)_2Cl_2 \xrightarrow{2H_2O} HO \rightarrow Si(Cl)_2(OH_2)_2 \leftarrow OH \xrightarrow{-2HCl} HO \rightarrow Si(OH)_2 \leftarrow OH$$

Silicic acid

$$\longrightarrow SiO_2 + 2H_2O$$

Elements with an electron sextet: Examples are oxygen and sulphur

$:\ddot{O}$	+	$:SO_3^{2-}$	$\rightarrow$	$(O \leftarrow SO_3)^{2-}$
Lewis acid		Lewis base		Adduct (Sulphate)
$:\ddot{S}$	+	$:SO_3^{2-}$	$\rightarrow$	$(S \leftarrow SO_3)^{2-}$
				Thiosulphate

4. *Molecules with multiple bonds between dissimilar atoms*. Typical examples of this class are CO_2, SO_2 and SO_3. In these compounds, the oxygen atoms are more electronegative than carbon or sulphur. As a result, the electrons are displaced away from the carbon or sulphur atom towards oxygen atoms. The carbon or sulphur atom thus becomes electron-deficient and can form dative bond with Lewis base, such as OH^-. $\overset{\delta^-}{:\ddot{O}}=\overset{\delta^+}{C}=\overset{\delta^-}{\ddot{O}:} + :\ddot{O}H^- \rightarrow \left[\text{O}{=}\text{C}(\text{OH}){=}\text{O} \right]^{-1}$

Lewis bases include all anions; molecules having unshared pair of electrons such as water, alcohols, ether; and compounds like carbon monoxide, nitric oxide etc. which can form the complexes with metals.

CLASSIFICATION OF LEWIS ACID-BASE REACTIONS

Lewis acids are also called electron acceptors, and Lewis bases, electron donors. The reaction product of the two may be called an adduct, an addition compound, a coordination complex, or an acid-base complex.

$$BF_3 + :NH_3 \rightleftharpoons F_3B{:}NH_3 \quad (1)$$

Electron acceptor or acid		Electron donor or base		Adduct

The adducts formed by neutralization reactions can function, in turn, as Lewis acids or bases toward other substances. They also can undergo various displacement reactions in which the original acid-base components of the adduct are replaced by another Lewis acid or bases or are interchanged with the components of another adduct.

It can be seen that Bronsted acids are actually adducts of the Lewis acid, H^+. The more stable the Lewis adduct, the weaker it is as a

Bronsted acid and, conversely, the weaker the adduct, the stronger it is as a Bronsted acid. Also, if the solvation of the ions is ignored, all neutralizations of the solvent cation and anion in the solvent system, as well as all interactions of a cation with an anion, can be viewed as being either direct Lewis neutralization reactions or displacement reactions. For example:

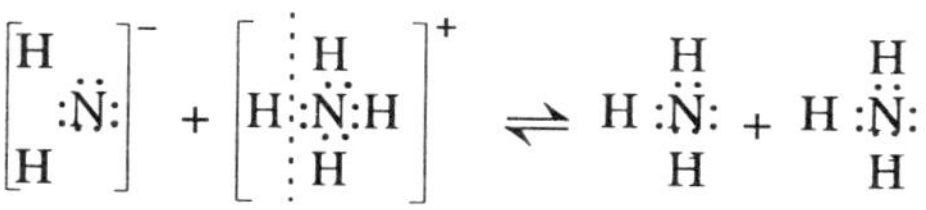

$$\left[H:\ddot{N}:H\right]^- + \left[H:\ddot{N}:H \atop H\right]^+ \rightleftharpoons H:\ddot{N}:H + H:\ddot{N}:H$$

Base$_1$ Acid Base$_2$

$$Ag^+ + [:Cl:] \rightleftharpoons Ag[:Cl:](s)$$

Acid Base

The interpretation of chemical reactions in terms of the Lewis acid-base definitions can be extended further to include the following phenomena:

Solubility interactions in which a complex of definite stoichiometry is formed, for example, in the dissolution of AlF_3 in anhydrous liquid HF,

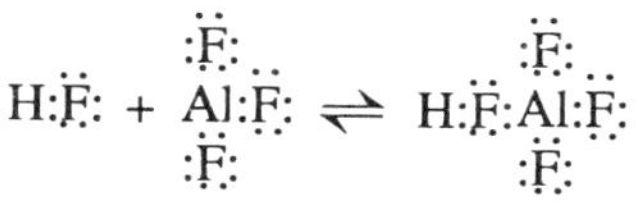

$$H:\ddot{F}: + Al:\ddot{F}:(:\ddot{F}:)_2 \rightleftharpoons H:\ddot{F}:Al:\ddot{F}:(:\ddot{F}:)_2$$

Base Acid

Ionic dissociation in a solvent,

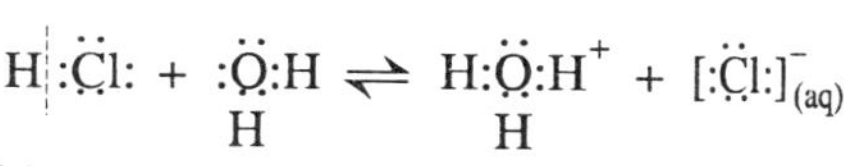

$$H:\ddot{Cl}: + :\ddot{O}:H(H) \rightleftharpoons H:\ddot{O}:H^+(H) + [:\ddot{Cl}:]^-_{(aq)}$$

Acid Base$_1$ Base$_2$

$$Mg^{+2}[:\ddot{Cl}:^-]_2 + 6:O:H(H) \rightleftharpoons Mg(:\ddot{O}:H(H))_6^{+2} + 2[:\ddot{Cl}:]^-_{(aq)}$$

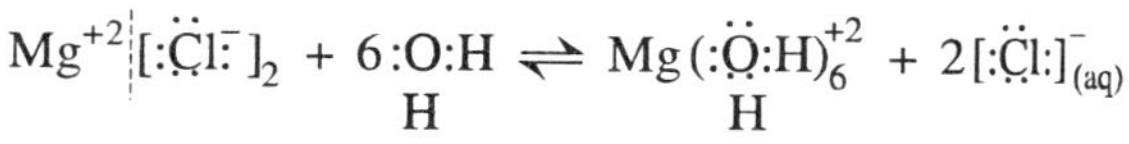

Acid Base$_1$ Base$_2$

Solvolysis reaction

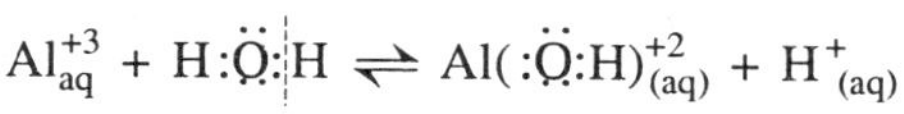

$$Al^{+3}_{aq} + H:\ddot{O}:H \rightleftharpoons Al(:\ddot{O}:H)^{+2}_{(aq)} + H^+_{(aq)}$$

Acid$_1$ Base Acid$_2$

$$[CN]^-_{(aq)} + H\vdots:\ddot{\underset{..}{O}}:H \rightleftharpoons H:CN + [:\ddot{\underset{..}{O}}:H]^-_{(aq)}$$

$\text{Base}_1 \quad \text{Acid}_1 \; \text{Base}_2$

The formation of metal complexes,

$$Ag^+ + 2:\overset{H}{\underset{H:}{\dot{N}}}:H \rightleftharpoons Ag(:NH_3)_2^+$$

Acid Base

$$Cu+2 + 6[:CN]^- \rightleftharpoons Cu(:CN)_6^{-3}$$

Acid Base

In a dissociation reaction the solvent molecule is able to act as an acid or a base, and is able to cause a displacement reaction in the solute adduct, giving as products charged adducts called solvated ions. In a solvolysis reaction, on the other hand, the solute adduct or, more commonly, one of the products of its dissociation, is a strong enough Lewis acid or base to cause a displacement reaction in the solvent itself. Still other common examples of solvolysis are the reactions of the so-called acidic and basic anhydrides with water:

$$\begin{matrix} :\ddot{O}: \\ :\ddot{O}:S \\ :\ddot{O}: \end{matrix} + \begin{matrix} :\ddot{O}:H \\ H \end{matrix} \rightleftharpoons \left[\begin{matrix} :\ddot{O}: \\ :\ddot{O}:S:\ddot{O}:H \\ :\ddot{O}: \end{matrix}\right]^- + H^+$$

$\text{Acid}_1 \; \text{Base} \; \text{Acid}_2$
(Acid anhydride)

$$Ca^{+2}[:\ddot{O}:]^{-2} + H\vdots:\ddot{O}:H \rightleftharpoons Ca^{+2} + 2[:\ddot{O}:H]^-$$

$\text{Base}_1 \qquad \text{acid}_1 \; \text{Base}_1$
(Basic anhydride)

In the case of metal complex formation the Lewis base is usually called a ligand. When complex formation occurs in water or some other solvent, the reaction really is a Lewis base displacement in which the ligand competes with the solvent for the metal cation or Lewis acid. The reaction between Co^{+3} and CN^- in water actually should be written:

$$Co \mid (H_2O)_6^{+2} + 6CN^- \rightleftharpoons Co(CN)_6^{-3} + 6HO$$

$\text{Acid} \mid \; \text{Base}_1 \quad \text{Base}_2$

Special cases of these phenomena are also classified as acid-base reactions according to the Arrhenius, Bronsted-Lowry, and solvent system definitions, but only the Lewis definitions are able to correlate all of them.

As seen from the above examples, the solvent itself often plays the role of an acid or base when reactions occur in solution. In fact, simple neutralizations, like Equation 1, seldom occur except in the gas phase or in very inert solvents. When we consider the effects of the solvent for reactions taking place in solution, including both the solvation of the anion as well as the cation for charged species (not illustrated in examples), most of the reactions take the form:

$$A_1B_1 + A_2B_2 \rightleftharpoons A_1B_2 + A_2B_1 \qquad (2)$$

where the solvent functions as either A_2 or B_1, or both. For example:

$$Ag(H_2O)_x^2 + (H_2O)_yCl^- \rightleftharpoons AgCl(s) + \frac{x+y}{2}(H_2O--H_2O)$$

Water and other polar solvents are able to play this joint role because they have both acidic and basic sites on their molecules. The positive end of any polar molecule is electron deficient and, hence, behaves as a weak Lewis acid. The negative end is electron rich and behaves like a weak Lewis base. Water is able to act as a base via the lone pairs on its oxygen atom and as an acid via hydrogen bonding. Though most of these solute-solvent interactions are ion-dipole or dipole-dipole in nature and weaker than most chemical bonds, they are, nevertheless, important in looking at the total acid-base picture in solutions.

Usanovich concept

A very comprehensive definition of acids and bases was proposed by Russian chemist Usanovich (1939). According to him: An acid is any chemical species which reacts with a base, gives up cations, or accepts anions or electrons. Conversely, a base is any chemical species which reacts with acids, gives up anions or electrons, or combines with cations. The Usanovich's definition includes all Lewis acids and bases. It also includes redox reactions which may consist of complete transfer of one or more electrons. For example acid SO_3 combines with anion O^{2-} and base Na_2O gives up anion O^{2-},

$$\underset{\text{acid}}{SO_3} + \underset{\text{base}}{Na_2O} \rightarrow Na_2SO_4$$

Other examples are: acid Cl gains an electron; base Na loses an electron:

$$Cl_2 + 2Na \rightarrow 2NaCl$$

acid $Fe(CN)_2$ combines with CN^- anions; base KCN gives up CN^-:

$$Fe(CN)_2 + 4KCN \rightarrow K_4Fe(CN)_6$$

acid Sb_2S_5 combines with anion S^{2-}; base $(NH_4)_2S$ gives up anion S^{2-}:

$$Sb_2S_5 + 3(NH_4)_2S \rightarrow 2(NH_4)_3SbS_4$$

It is apparent that Usanovich concept is a synthesis of all the previous acid-base theories and in the process it has become too general. A comparison of the several concepts of acid-base reactions reveals that it is difficult to make an appropriate choice for the best.

STRENGTH OF ACIDS

According to Bronsted-Lowry concept, a strong acid is one which has a strong tendency to donate a proton. A comparison of acid strength is generally done by comparing proton-donating tendencies of two acids to the same base; generally water is used as the base. For an acid HA donating a proton to the base H_2O we may write

$$HA + H_2O \rightleftharpoons H_3O^+ + A^-$$

so that

$$K_a = \frac{[H_3O^+][A^-]}{HA}$$

measures the tendency of the acid HA to protonate H_2O. It is obvious that larger the value of HA the stronger the acid and vice versa. Thus if we compare the relative acidic strengths of CH_3COOH ($K_a = 1.8 \times 10^{-5}$) and HCN ($K_a = 7.2 \times 10^{-10}$), the former is a stronger acid.

For a base B we may similarly write the equilibrium reaction

$$B + H_2O \rightleftharpoons BH^+ + OH^-$$

and the equilibrium constant

$$K_b = \frac{[BH^+][OH^-]}{[B]}$$

K_b measures the strength of the base. B is a stronger base than OH^- if it can extract a proton from H_2O; larger the value of K_b the stronger is the base.

It is generally more convenient to express K_a and K_b values in terms of their negative logarithms i.e. as pK_a (which is = – log K_a) and pK_b. Then, a small value of pK_a means the acid is stronger, similarly small value of pK_b indicates a stronger base.

STRENGTH OF HYDRACIDS

The correlation between acid or base strength is complex and involves many factors. Some of these are elaborated as follows:

In such polar molecules with a hydrogen atom situated at one end of the dipole, electrons are withdrawn from the hydrogen atom, thereby facilitating its release as a proton. Thus the acid strength should increase with increase in electronegativity of the atom combined with hydrogen. For the hydracids of group VIA (16) and VIIIA (17), the relative strengths are in the order

$$H_2Te > H_2Se > H_2S > H_2O$$

and

$$HI > HBr > HCl > HF$$

The acid strength seems to decrease with increasing electronegativity by the least electronegative element of the respective group. Evidently other factors are also involved:

(a) *Stability of the conjugate base*: Charge density on the conjugate bases is in the order

$$F^- > Cl^- > Br^- > I^- \text{ and } O^{2-} > S^{2-} > Se^{2-} > Te^{2-}$$

Greater charge density on the conjugate base will result in greater proton attraction by the conjugate base. The corresponding acid will thus be weaker.

Moreover a large anion will be more stabilized by solvation leading to more ionization of the acid. Hence, smaller anion results in the formation of the weaker hydracid.

(b) *Hydrogen-bonding*: Smaller anions are bound by hydrogen bonding also, decreasing the acid strength of the hydrogen compounds.

(c) *H-X bond energy*: Decrease in the bond strength of H-X bonds parallels the decrease in pK_a and, hence, increases the acid strength.

For a better understanding of the variation of acid strength of hydracids, a Born-Haber cycle of the type shown below should be considered.

$$\begin{array}{ccc}
HX(g) & \xrightarrow{+D} & H(g) \;+\; X(g) \\
\uparrow & & +I \downarrow \qquad \downarrow -EA \\
| & & H^{+}(g) \;+\; X^{-}(g) \\
| & & \Delta H_{(hyd)_1} \downarrow \qquad \downarrow \Delta H_{(hyd)_2} \\
HX(a) & \xrightarrow{\Delta H} & H^{+}(aq) \;+\; X^{-}(aq)
\end{array}$$

where D is the dissociation energy of gaseous HX; ΔH the enthalpy of $H_{\text{(hyd)}}$ is the enthalpy of ionization of the ions; $H_{(evap)}$ is the enthalpy of evaporation of HX solution. Thus

$$\Delta H = \Delta H\,(\text{hyd})_1 + \Delta H\,(hyd)_2 + \Delta H\,(evap) + D + I - EA$$

Strength of Oxyacids

Oxyacids are derived from oxyanions by simply adding one or more protons to the ion:

$$n\text{H}^{+} + \text{XO}_m^{n-} \rightarrow \text{H}_n\text{XO}_m$$

(where X = B, Si, Ge, N, P, As, S, Se, F, Cl, Br or I)

When the oxyacid is dissolved in water, the protons, normally bonded to the oxygen atoms, may be transferred from the acid to the solvent (water). This transfer is due to competition between the water and the oxyanion as proton acceptors. An oxyacid which is a good proton donor is termed as a strong acid; one that is a poor proton donor is a weak acid. In either case, an equilibrium involving the acid and the water is established:

$$\text{H}_2\text{O} + \text{H}_n\text{XO}_m \rightleftharpoons \text{H}_3\text{O}^{+} + \text{H}_{n-1}\ \text{XO}_m^{-}$$

The extent to which protons are transferred to water may be correlated with the structure of the oxyanion and with the ionic potential of the hypothetical central ion, X^{n+}, where n is the oxidation state of the central element. In the simple oxyacid:

$$\text{X} - \text{O} - \text{H}$$

cleavage of the X-O bond in aqueous solution is characteristic of a base, and cleavage of the O-H bond is characteristic of an acid. When either of the two bonds can be cleaved by an appropriate reagent, the substance is amphoteric. Here, we are concerned with the acidic behavior or cleavage of the O-H bond. To find the reason for the wide variations in ionization constants among the oxyacids the following general rules are helpful.

Rule 1. The O-H bond in X-O-H is easily cleaved whenever X has a high ionic potential. When the ionic potential (ratio of charge to size) is high bonding electrons are displaced toward X and away from hydrogen in the X-O-H structure, and hence the acid strength of the oxyacid is greater.

Oxyacids of a given Element

For oxyacids of a given element, the higher the oxidation number of the hypothetical central ion, the higher its ionic potential. Consequently, when X is chlorine, $HClO_4$ is the strongest acid in the series $HClO_3$, $HClO_2$, $HClO_2$ and HClO, while HClO is the weakest acid. Similarly, the strengths of the oxyacids of nitrogen and sulphur are in the orders:

$$HNO_3 > HNO_2$$
$$H_2SO_4 > H_2SO_3$$

In these comparisons we concern ourselves with the loss of only the first proton by the acid, further dissociation being too small.

Oxyacids of Elements in a Group

For a series of oxyacids derived from elements within a group, in which the central atoms have identical oxidation states, the first member is the strongest acid. For example, in the series $HClO_3$, $HBrO_3$, and HIO_3, chloric acid is the strongest acid and iodic acid is the weakest. Other series of oxyacids within groups and their relative acid strengths are:

$$HOCl > HOBr > HOI$$
$$H_2SO_4 > H_2SeO_4 > H_2TeO_4$$
$$H_2SO_3 > H_2SeO_3 > H_2TeO_3$$
$$HNO_3 > H_3PO_4 > H_3AsO_4$$

The differences in strength of these oxyacids may be attributed to the differences in ionic potentials of the central atoms. As we move down a group, the size of the central atom increases, and the ratio of charge to size decreases, resulting in less ready dissociation of the O-H bonds in the oxyacid.

Oxyacids of Elements in a Period

The strengths of oxyacids formed by elements in a given period may be predicted from examination of the ionic potentials of the central atoms, provided the central atom is in its maximum oxidation state. Examples of this trend are

$$HClO_4 > H_2SO_4 > H_3PO_4 > H_4SiO_4$$
$$HNO_3 > H_2CO_3 > H_3BO_3$$
$$H_2SeO_4 > H_2AsO_4 > H_4GeO_4$$

Because the charge of the hypothetical central ion increases from left to right in a period and size decreases, those oxyacids derived from Group VII elements are the most highly dissociated in an aqueous solution.

For strong acids, relative acid strengths cannot be determined in aqueous solutions, since water is so effective as a proton acceptor. In fact water is termed a '**leveling**' solvent for strong acids. Relative acid strengths of strong oxyacids, $HClO_4$, HNO_3, H_2SO_4, HIO_3 etc., can be measured in non-aqueous medium.

Rule 2. Another rule for predicting the relative strength of an oxyacid is based on the ratio of the oxidation number of the central element to the number of atoms bonded to that element: the higher this ratio, the stronger the acid. In H_4SiO_4 or $Si(OH)_4$, four atoms are bonded to the silicon atom, which has an oxidation state of IV. Thus the ratio is 4/4, or 1. In H_3PO_4 or $PO(OH)_3$, there are four atoms bonded to the phosphorus atom, but its oxidation number is five, therefore the ratio is 5/4. For H_2SO_4 the ratio is 6/4; for $HClO_4$, 7/4. As we move through this series of oxyacids, each oxygen atom is left with less and less bonding power for protons, and hence the acid becomes stronger.

Rule 3. A useful rule of thumb for qualitative distinction between weak and strong acids is derived by applying the principle of electroneutrality to the corresponding oxyanion.

(i) Oxyanions that have an ionic charge of less than $-\frac{1}{2}$ per oxygen atom are observed to form strong acids;

(ii) those with a charge of more than $-\frac{1}{2}$ per oxygen atom form weak acids.

(iii) those with exactly a charge of $-\frac{1}{2}$ per oxygen atom form acids that may be weak or strong.

For example, CO_3^{2-}, PO_4^{3-}, SO_3^{2-}, and ClO^- all have an ionic charge per oxygen atom, greater than $-\frac{1}{2}$ and form weak acids. The ions, NO_3^-, ClO_3^-, MnO_4^- have an ionic charge per oxygen atom of less than $-\frac{1}{2}$ and form strong acids. SO_4^{2-}, SeO_4^{2-}, ClO_2^-, and NO_2^- have an ionic charge of exactly $-\frac{1}{2}$ per oxygen atom; SO_4^{2-} and SeO_4^{2-} form strong and ClO_2^- and NO_2^- form weak acids.

The rule cannot be applied to meta-oxyanions with corresponding ortho-forms since the meta-form is converted to ortho-in aqueous solution where determination of acid strength is carried out.

For example $(PO_3^-)_n$ metaphosphate ion having a charge $-\frac{1}{3}$ per oxygen atom should form a strong acid; but in aqueous solution it is converted to orthophosphate, PO_4^{3-} with a charge of $-\frac{3}{4}$ per oxygen.

The rules discussed in the foregoing discussion may be summarised as follows:

1. For a given element increase in the oxidation number of the central atom X increase the acidity.

	HOCl <	HOClO <	$HOClO_2$ <	$HOClO_3$
oxidation number	+1	+3	+5	+7

2. For oxyacids within a periodic group having identical oxidation number, the acid strength decreases with increase in size of the central atom *X*. In other words, the acid strength increases with increase in electron-negativity of the central atom X due to inductive effect. For example,

	Oxidation number
HOCl > HOBr > HOI	+1
$HClO_3$ > $HBrO_3$ > HIO_3	+5
HNO_3 > HPO_3 > $HAsO_3$	+5
H_2SO_4 > H_2SeO_4 > H_2TeO_4	+6

3. The relative strength of an oxyacid can be predicted on the basis of the ratio of the oxidation number of the central atom to the number of atoms bonded to that element, the higher ratio, the stronger the acid. Thus, in the series

$$H_2SiO_4 < H_3PO_4 < H_2SO_4 < HClO_4$$

or $Si(OH)_4 < OP(OH)_3 < O_2S(OH)_2 < O_3Cl(OH)$

or $Si(OH)_4$ $OP(OH)_3 < O_2S(OH)_2$ $O_3Cl(OH)$

Ratio 4/4 5/4 6/4 7/4

as the ratio increases the acids become stronger.

4. Acid strength may be correlated to the resultant charge on oxygen atoms applying electroneutrality principle. If the oxyanions have a charge $> -\frac{1}{2}$ per oxygen atom the acid is weaker and if $< -\frac{1}{2}$ it is strong. In case the charge $= -\frac{1}{2}$ the acid may be weak or strong.

PAULING'S RULES

Regularities in the observed strength of oxyacids have also been formulated in terms of two rules by Pauling:

Rule 1. For polybasic mononuclear oxyacids, successive dissociation constants K_1, K_2, K_3, ... are in the ratios $1: 10^{-5}: 10^{-10}: ...$

Rule 2. The value of the first dissociation constant (K_1) is determined by the value of *m* in the formula $OX_m (OH)_n$, if *m* is zero [no excess of oxygen atoms over hydrogen atoms, as in $B(OH)_3$] the acid is very weak, with $K_1 \leq 10^{-7}$, for $m = 1$, the acid is weak, with $K_1 \simeq 10^{-2}$, for $m = 2$ ($K_1 = 10^3$) or $m = 3$ ($K_1 = 10^8$) the acid is strong.

The first rule is illustrated by the following example:

$$H_3PO_4 \rightleftharpoons H^+ + H_2PO_4^-; \quad K_1 = \frac{[H^+][H_2PO_4^-]}{[H_3PO_4]} = 7 \times 10^{-3}$$

$$H_2PO_4^- \rightleftharpoons H^+ + HPO_4^{2-}; \quad K_2 = \frac{[H^+][HPO_4^{2-}]}{[H_2PO_4^-]} = 6.3 \times 10^{-8}$$

$$HPO_4^{2-} \rightleftharpoons H^+ + PO_4^{3-}; \quad K_3 = \frac{[H^+][PO_4^{3-}]}{[HPO_4^{2-}]} = 4.2 \times 10^{-13}$$

The second rule may be clearly understood as follows. Consider the acid XOH, the force attracting H^+ to XO^- to form XOH is that of an O-H valence bond. But the force between H+ and either one of the two O atoms in XO_2^- to form XOOH may be smallei than that for an O-H bond because the total attraction for the proton is divided between the two O atoms. Thus, this acid may be expected to be more highly dissociated than XOH.

This rule is illustrated by examples in Table 2.2.

Table 2.2 Values of pKa for some oxyacids of general composition $XO_m(OH)_n$

$X(OH)_n$ (very weak)		$XO(OH)_n$ (weak)		$XO_2(OH)_n$ (strong)		$XO_3(OH)_n$ (very strong)
		NO(OH)	3.3	$NO_2(OH)$	−1.4	–
Cl(OH)	7.2	ClO(OH)	2.0	$ClO_2(OH)$	−1	$ClO_3(OH)$ −10
Br(OH)	8.7	–		–		–
I(OH)	10.0	–		$I0_2(OH)$	0.8	–
$B(OH)_3$	9.2	$CO(OH)_2$	3.9	–		–
–		$SO(OH)_3$	1.9	$SO_2(OH)_2$	0	–
$As(OH)_3$	9.2	$PO(OH)_3$	2.1	–		–
$Si(OH)_4$	10.0	$IO(OH)_5$	1.6	–		–

ELECTRONEGATIVITY AND THE ACID-BASE CHARACTER OF BINARY OXIDES

The Pauling electronegativity scale can be used to predict standard enthalpies of formation of binary compounds at 25 °C via the equation

$$\Delta H_f^\circ = -96.5 \cdot n \cdot [X_A - X_B]^2 \quad (1)$$

In eq 1, ΔH_f° is the standard enthalpy of formation in kilo-joules per mole, n is the number of equivalents in the compound formula, and X_A and X_B are the Pauling electronegativities for the two elements under consideration. Pauling's electronegatives have units of (electron volts)$^{1/2}$; the conversion factor between electron volts and kilojoules per mole is 96.5 (the Faraday constant). As discussed by Allred, eq 1 is only reliable within about ±20% and fails for compounds where $[X_A - X_B]$ is greater than about 1.8 $(eV)^{1/2}$. Still, it is remarkably successful in view of the fact that no consideration is given to structural changes or physical states of reactants or products.

Smith has recently compiled a numerical scale of acid-base character for binary oxides by analogy with the Pauling electronegativity scale. For each oxide, he has evaluated an acid-base parameter *a*, which can be used to predict standard enthalpies of combination at 25 °C of binary oxides to form oxo-salts via the equation

$$\Delta H^\circ_{comb} = -[a(A) - a(B)]^2 \qquad (2)$$

In eq 2, ΔH°_{comb} is the standard enthalpy change in kilojoules per mole of oxide ion transferred from basic oxide B to acidic oxide A, and $a(A)$ and $a(B)$ are Smith's acid-base parameters for the two oxides under consideration. Smith has arbitrarily referenced his acid-base parameters to $a(H_2O) = 0$; oxides with $a < 0$ tend to be basic or amphoteric, while oxides with $a > 0$ tend to be acidic. Smith's acid-base parameters have units of $(\text{kilojoules/mole})^{1/2}$. Equation 2 is more reliable than eq 1, generally well within ±10%, perhaps because it is more specific in its application.

It is of interest to investigate Smith's acid-base parameters within

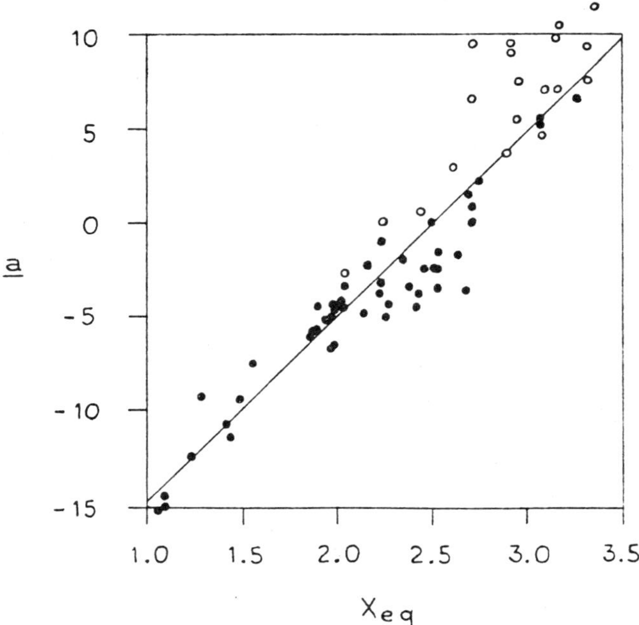

Plot of Smith's acid–base parameters *a* versus equalized electronegativities X_{eq} (eq 3). Pauling electronegativities from ref 1 and 2. The line drawn corresponds to eq 5.

the framework of the Pauling electronegativity system and modern electronegativity theory. Sanderson's principle of electronegativity equalization states that two or more atoms of different initial (prebonded) electronegativities equalize their electronegativities on combination. It has been suggested that the equalized electronegativity X_{eq} in an oxide may be estimated by

$$X_{eq} = \frac{N}{\Sigma(v/X)} \tag{3}$$

In eq 3, X_{eq} is the equalized electronegativity of the oxide, N is the number of atoms in the oxide formula, v is the number of atoms of each element in the oxide formula, and X is the initial, prebonded electronegativity of each element on the Pauling scale.

A possible relationship between Smith's oxide acid-base scale and the Pauling electronegativity scale is

$$a = m \cdot X_{eq} + b \tag{4}$$

where m and b are a slope and an intercept, respectively. Considering the units of the Smith and Pauling scales as given above, the theoretical slope m should be the square root of the Faraday constant, 9.8 [kJ (mol. eV)]$^{1/2}$. The theoretical intercept b should be that which gives $a(H_2O) = 0$; using $X_{eq}(H_2O) = 2.50$ (eq 3), $b = -24.5$ (kJ/mol)$^{1/2}$. Thus, eq 4 can be rewritten as

$$a = 9.8 \cdot X_{eq} - 24.5 \tag{5}$$

A plot of Smith's acid-base parameters a versus equalized electronegativity X_{eq} (eq 3) is provided in the figure. Solid points represent oxides for which the electronegativity of the other element is available for the oxidation state under consideration; open points represent oxides for which it is not. The line drawn corresponds to eq 5 and is *not* a statistically derived best fit. It is seen that the solid points obey eq 5 reasonably well; deviations from the line drawn may be attributed to uncertainties in both X_{eq} and a.

Open points in the figure represent oxides for which the electronegativity of the other element is not available for the oxidation state under consideration. For example, the electronegativity of manganese (VII) is unavailable, and Mn_2O_7 is represented in the figure by an open point, estimating X_{eq} $(Mn_2O_7) = 2.71$ by using $X_{Mn} = 1.55$, which is the value given by Allred for manganese (II). Because the electronegativity of manganese (VII) is probably much greater than 2.71, and the point for Mn_2O_7 should be displaced toward the right i.e., closer to the line given by eq 5. A similar argument applies to other oxides represented in the figure by open points.

Relationship between Standard Enthalpies of Combination of Binary Oxides and Electronegativity

Combination of eqs 2 and 5 gives

$$\Delta H^{\circ}_{comb} = -\ 96.5 \cdot [X_{eq}(A) - X_{eq}(B)]^2 \tag{6}$$

Equation 6 can be used to estimate standard enthalpies of combination of binary oxides (in kilojoules per mole of oxide ion transferred) in cases where experimentally determined acid-base parameters are unavailable. Equation 6 is about as reliable as eq 1, i.e. within about ±20%. It is significant that eq 6 requires no adjustable parameters, thereby providing support for the calculation of X_{eq} via eq 3.

Relationship between Acid-Base Character of Binary Oxides and Partial Charge on Combined Oxygen

The partial charge on combined oxygen may be estimated by

$$\delta_0 = \frac{X_{eq} - X_0}{X_0} \tag{7}$$

In eq 7, δ_0 is the partial charge on combined oxygen, X_{eq} is the equalized electronegativity of the oxide (eq 3), and X_0 is the initial, prebonded electronegativity of oxygen, 3.44.

Solving eqs 5 and 7 for X_{eq} and setting them equal to one another gives

$$\frac{a + 24.5}{9.8} = X_{eq} = X_0 \cdot \delta_0 + X_0 = 3.44 \cdot \delta_0 + 3.44$$

or

$$a = 33.7 \cdot \delta_0 + 9.2 \tag{8}$$

Sanderson has qualitatively discussed the relationship between oxide acid-base character and partial charge on combined oxygen, noting that oxides with very negative δ_0 tend to be basic, oxides with moderately negative δ_0 tend to be amphoteric, and oxides with only slightly negative δ_0 tend to be acidic. Equation 8 puts Sanderson's ideas on a quantitative basis.

Strength of Organic Acids

Organic acids may be represented in general as RCOOH i.e. those having a carboxylic acid group and R-OH i.e. those having an OH group capable of releasing a proton on ionization. The acid strength depends on the extent of ionization or release of H^+. The extent of ionization or the degree of acidity may be determined by:

(*i*) *The electronegativity of the group attached to H*: The greater the electronegativity the stronger the acid due to inductive effect. For example consider CH_3O–H and H_3C–H; electronegativity of O being considerably greater than that of C, CH_3OH should be a stronger acid. This is actually found to be the case

$$\begin{array}{lcc} & CH_3OH & CH_4 \\ pK_a & \simeq 16 & \simeq 50 \end{array}$$

(*ii*) *Resonance effects*: An important factor is the stabilization of the anion of an acid through resonance effect. For example for the formate ion two structures are:

$$\begin{array}{ccc} \mathrm{O} & & \mathrm{O^-} \\ \| & & | \\ \mathrm{H{-}C{-}O^-} & \leftrightarrow & \mathrm{H{-}C{=}O} \end{array}$$

(two canonical structures of identical energy)

Effect of Substituents on the Strength of Acids

(1) *Resonance effects*: Phenol is a stronger acid than alcohol, since phenoxide ion $C_6H_5O^-$ is stabilized by resonance whereas no such stabilization occurs in alkoxide ion. Any factor that stabilizes the anion should increase the acidity. Similarly factors which destabilise the anion should decrease the acidity.

In nitrophenols the nitro group is an electron withdrawing group. Due to the inductive effect, it will increase the acidity of phenol in the order ortho > meta > para. But due to the resonance, weakening of O-H bond shall be maximum when the nitro group is in para position. Thus the acidity of nitrophenols follows the order

para > ortho > meta

(2) *Inductive effect*: An electron-withdrawing substituent stabilizes the anion by dispersing the negative charge and, therefore, increase the acidity. On the other hand, electron-releasing substituents increase the negative charge on the anion thus destabilising it. Consequently acidity of the acid decreases. Let us consider substituted acetic acids. In case of XCH_2COOH, the element X is an electron withdrawing group; thus, the carbon atom attached to X gets a slight positive charge. This carbon, in turn, withdraws electrons from the C=O and C—O—H bonds. The resultant effect is weakening of the O-H bond and the dissociation of a proton becomes easier. The substituted acid XCH_2COOH thus becomes stronger than CH_3COOH.

If the substituent group is an electron releasing the group such as CH_3, the electron density on the carbon atom increases which in turn increases electron density on the adjoining C=O and C-O-H bonds. The overall effect is increase of electron density in the O-H bond. Consequently the release of proton becomes difficult . Thus the acidity decreases.

$$\overset{\delta^-}{X} \leftarrow \underset{H}{\overset{H}{\underset{|}{\overset{|}{C}}}} \leftarrow \overset{O}{\overset{\|}{C}} \leftarrow O \leftarrow \overset{\delta^+}{H} \qquad \overset{\delta^+}{CH_3} \rightarrow \underset{H}{\overset{H}{\underset{|}{\overset{|}{C}}}} \rightarrow \overset{O}{\overset{\|}{C}} \rightarrow O \rightarrow \overset{\delta^-}{H}$$

Electron-withdrawing group — + I effect

Electron- releasing group — – I effect

Strength of Bases

A base must have an unshared electron pair in order to attract and hold a proton; a polar molecule that has an electron pair in order to attract and hold a proton; a polar molecule that has an electron pair situated at the negative end of the dipole is a base.

Trends in the base strength of anions are readily derived from conjugate relationships. Thus, since H_2S is a stronger acid than H_2O, S^{2-} is a weaker base than O^{2-}.

For monoatomic anions of similar charge, base strength decreases with increasing size of the atom. The base strength of the molecules, NH_3, H_2O and HF increases with a decrease in electronegativity of the element. Thus the order is

$$NH_3 > H_2O > HF$$

A similar order is observed for the uninegative ions of these molecules:

$$NH_2^- > OH^- > F^-$$

In these series, size differences are small and unimportant; the trend is set by the decreasing electronegativity of the central element. Both small size and low electronegativity are found in the hydride ion, H^-, which is a powerful base.

Another factor that influences the base strength of anions is the charge on the ion. Thus, the base strength of the monoatomic anions of the elements of the second period

$$N^{2-} > O^{2-} > F^-$$

decreases with increasing electronegativity and with decreasing negative charge on the ion. To summarize:

The base strength decreases with (*i*) increasing size of the atom, (*ii*) increasing electronegativity of the element and (*iii*) decreasing negative charge on the ion.

Relative Strengths of Lewis Acids and Bases

Methods that have been used for comparing the strengths of a series of Lewis acids or bases include the following:

(i) Gas-phase dissociation data,
(ii) Calorimetric heats of reactions,
(iii) Competitive acid-base reactions,
(iv) Displacement reactions.

In general, the strengths of Lewis base towards Lewis acids parallel their basicity towards a proton, provided that only π bonds are formed in adduct formation and steric factors are not important.

The following trends are observed:

1. *Position in the Periodic Table*: In general, one might expect Lewis base strength to decrease as a group is descended. Similarly, base strength is found to decrease from Me_3N to Me_2O and then to MeF. Me_3N displaces both Me_2O and MeF, whereas Me_2O will displace only MeF from Me_3B adducts. However, caution must be exercised in making generalisations.

2. *Effect of substituents*: The strength of Lewis acids and bases can

often be predicted by considering the inductive, resonance and steric effects of the substituents. For example, replacing hydrogen atoms by an electron-withdrawing group will increase the acid strength of a Lewis acid and decreases the base-strength of a Lewis base. Thus, acid-strength increases in the series $Me_3B > BH_3 > BF_3$, whereas base-strength decreases in the series $Me_3N > NH_3 > NF_3$. These predictions have experimentally been found to be correct. The experimental order of acid-strength. $BF_3 < BC1_3 < BBr_3$ is, however, not the order expected from inductive effects. This is due to resonance stabilization or $P\pi$–$P\pi$ back bonding which decreases from BF_3 to BBr_3.

Ammonia: NH_3 is a weak base in aqueous medium but aqueous nitrogen trifluoride (NFa) is not at all basic. This is due to the inductive effects of substituents on nitrogen. Nitrogen is partially positive in NFa due to higher electronegativity of F. Replacement of H in NH_3 by any electron withdrawing group like —CF_3, —NH_2, —OH, —OR, etc. reduces the basicity. On the other hand, replacement of H by electron releasing alkyl group increases the electron density on N and hence increases basicity $MeNH_2$, for instance, is more basic than NH_3. Replacement of second H to form Me_2NH further increases the basicity. In general, basic character of amines follows the order:

$$R_2NH > RNH_2 > NH_3$$

However, this increase in basicity does not go beyond secondary amines and it has been observed that tertiary amines R_3N are weaker bases with basicity of the order of primary amines or less. The reason for this anomaly is the increased steric effect of three alkyl groups on a small nitrogen atom and lower energy of solvation of tertiary amines as compared to other.

3. *Steric effects*: Steric considerations are seemingly unimportant in discussing strengths of simple protonic acids and bases, but they assume importance in more complicated systems. For example, 2-methyl pyridine is a stronger base than pyridine towards proton, but it is a weaker base than pyridine towards Lewis acids such as trimethyl boron, Me_3B. Similarly, trimethyl amine is a stronger base than ammonia towards proton but a weaker base towards $(C_4N_9)_3B$.

The steric hindrance such as the one seen in case of 2-methyl pyridine-BMe_3 complex is termed front or 'F-strain' and has a significant effect on the stability of the complex due to the bulk of the alkyl groups and their random rotational movement. The F-strain is so predominant that 2-tert-butylpyridine does not form a complex with BMe_3.

Another steric effect, the B-strain or back-strain, is observed in adducts in which bulky substituents are present on the donor or acceptor atoms. In the complexation of boron trialkyles, Br_3, with trialkylamines, NR_3^-, the donor N atom is required to be in sp^3 hybridised states so that the done pair can form strong bond. If the alkyl groups on the N atom are bulky, they force open the bond angle to open, and in extreme cases change the hybridisation from sp^3 to sp^2 state. The lone pair is then placed in a pure p-orbital which is poorly suited for coordination, such as in trisilylamine.

If, on the other hand, the R group in BR_3 are bulky, change of hybridisation from sp^2 to sp^3 causes strain on the molecule. The alkyl groups branch off reducing the Lewis acid strength. Thus, trimesityl boron, having bulky groups is almost inert to adduct formation with trialkylamines.

In amines, B-strain results from structural necessity at the nitrogen atom carrying very large substituents. Normally, amines (R_3N) are tetrahedral molecules (sp^3 hybridised N atoms) but if substituents are very large they force the bond angle to open up; in some cases the molecule may flatten up to give a plane triangular sp^2 hybridised molecule with the lone pair in a pure *p*-orbital. Such molecules are poor donors due to large B-strain because the large substituents will have to *fold back* behind the nitrogen atom to acquire sp^3 hybridization in the process of coordination with an acceptor.

Triethylamine, $(C_2H_5)_3N$, and quinuclidine are expected to have similar basicity and exhibit the same proton affinities. Models show no extra strain in quinuclidine relative to $(C_2H_5)_3N$. However, quinuclidine complex with Me_3B shows higher stability than Et_3N complex a large adduct bond energy (84 kJ mol^{-1}). Lower stability of Et_3N. BMe_3 is attributed to B-strain in Et_3N because in the process of formation of complex with BMe_3, the ethyl groups in Et_3N have to fold back (to avoid F-strain of methyl groups of BMe_3) which causes steric congestion between terminal methyls of ethyl groups (B-strain). On the other hand, alkyl groups in quinuclidine are already structured in the back position in the molecule with the result there is no B-strain on complexation.

4. *Effect of solvents*: Solvents play an important role in altering the strength of an acid or a base. Non-aqueous solvents such as ammonia,

amides ($RCONH_2$), hydrazine, etc. are *proton acceptors* and hence increase the ionisation of acids. Consequently, *in a basic solvent, all acids tend to become strong.* The reverse is true of acidic solvents such as CH_3COOH, hydrogen halides anhydrous sulphuric acid etc., *in which all bases tend to become strong* but *acids become relatively weak* or may even act as a base; for example nitric acid, a strong acid in aqueous medium, behaves as a base in anhydrous sulphuric acid

$$HNO_3 + 2H_2SO_4 \rightleftharpoons NO_2^+ + H_3O^+ + 2HSO_4^-$$

Such acidic solvents tend to exert differentiating effect on the strength of acids.

For example, HCl, HBr and HI which are all very strong acids in water (water exerting a levelling effect), act as weak electrolytes in acetic acid. Thus, acetic acid acts as a differentiating solvent towards halogen acids and their ionization increases with increases in atomic number of halogen.

That is, acid strength decreases as

$$HI > HBr > HCl.$$

Acetic acid has been extensively used to determine the relative strengths of strong acids like $HClO_4$, H_2SO_4, HNO_3, halogen acids, etc. The following trend has been noted:

$$HClO_4 > HI > HBr > HCl > H_2SO_4 > HNO_3.$$

However, acetic acid exerts a levelling effect on the strength of bases. Consequently, the relative strength of ammonia, primary amines, secondary amines, tertiary amines, etc. can not be distinguished, using acetic acid as a solvent.

Hard and Soft Acids and Bases (HSAB)

Ahrland, Chatt and Davies divided metal ions into two classes, viz., class (a) acceptors and class (b) acceptors. Metal ions which form their most stable complexes with ligands containing nitrogen, oxygen or fluorine as donor atom are termed class (a) acceptors and those which form their most stable complexes with ligands containing a heavier member of these families as the donor atom are termed class (b) acceptors.

Features which bring out class (a) acceptor are:

(1) Small size,

(2) High polarising power and

(3) High positive oxidation state.

Class (a) includes ions of alkali metals, alkaline earth metals, lighter transition metals in higher oxidation states such as Ti^{4+}, Cr^{3+}, Fe^{3+} and Co^{3+} and the hydrogen ion, H^+. Their outer electrons or orbitals are not easily distorted. The ligands which prefer to combine with these ions are also placed in class (a). For example, NH_3, R_3N, H_2O and F^- ion have great tendency to coordinate with class (a) metal ions (Be^{2+}, Co^{3+}, Ti^{4+}, etc) and thus they belong to this class. The tendency of the ligands to form complexes with class (a) metal ions is in the order:

$$F > Cl > Br > I$$

$$O >> S > Se > Te$$

$$N >> P > As > Sb$$

Class (b) acceptor behaviour is associated with the following features:

(1) Large size,

(2) Low or zero oxidation state, and

(3) Having easily distorted outer electrons.

Class (b) includes ions of the heavier transition metals and those in lower oxidation states such as Cu^+, Ag^+, Hg_2^{2+}, Pd^+, Pt^{2+} and Mg^{2+}. The ligands which exhibit preference towards these metal ions also belong to this category. Thus, ligands such as R_3P (phosphines) and R_2S (thioethers) prefer to combine with Pd^{2+}, Pt^+, and Hg^{2+}. The complexing ability of the ligands for the metal ions of class (b) is in the order

$$F < C1 < Br < I$$

$$O << S < Se \sim Te$$

$$N << P > As > Sb$$

Pearson defines a soft base as one in which the donor atom is highly polarizable, has low electronegativity and is easily oxidised or is associated with empty low-lying orbitals. A hard base is defined as one with the opposite properties, viz, the donor is of low polarizability and high electronegativity, is hard to reduce, and is associated with empty orbitals of high energy. Thus, metal ions of class (a) become hard acids and the ligands of this class become hard bases. On the other hand acids and the ligands of this class become hard bases. On the other hand, metal ions and the ligands of class (b) become soft acids and soft bases respectively. This classification is, however, not very rigid and there are borderline cases. A complete classification is given in Tables 2.3, 2.4 and 2.5.

Table 2.3 Classification of Hard and Soft Acids

Hard acids

H^+, Li^+, Na^+, K^+(Rb^+, Cs^+)

Be^{+2}, $Be(CH_3)_2$, Mg^{+2}, Ca^{+2}, Sr^{+2}(Ba^{+2})

Sc^{+3}, La^{+3}, Ce^{+4}, Gd^{+3}, Lu^{+3}, Th^{+4}, U^{+4}, UO_2^{+2}, Pu^{+4}

Ti^{+4}, Zr^{+4}, Hf^{+4}, VO^{+2}, Cr^{+3}, Cr^{+6}, MoO^{+3}, WO^{+4}, Mn^{+2}, Mn^{+7}, Fe^{+3}, Co^{+3}

BF_3, BCl_3, $B(OR)_3$, Al^{+3}, $Al(CH_3)_3$, $AlCl_3$, AlH_3, Ga^{+3}, Ib^{+3}

CO_2, RCO^+, NC^+, Si^{+4}, Sn^{+4}, CH_3Sn^{+3}, $(CH_3)_2Sn^{+2}$

N^{+3}, RPO_2^+, $ROPO_2^+$, As^{+3}

SO_3, RSO_2^+, $ROSO_2^+$

Cl^{+3}, Cl^{+7}, I^{+5}, I^{+7}

HX (hydrogen-bonding molecules)

Borderline acids

Fe^{+2}, Co^{+2}, Ni^{+2}, Cu^{+2}, Zn^{+2}

Rh^{+3}, Ir^{+3}, Ru^{+3}, Os^{+2}

$B(CH_3)_3$, GaH_3

R_3C^+, $C_6H_5^+$, Sn^{+2}, Pb^{+2}

NO^+, Sb^{+3}, Bi^{+3}

SO_2

Soft acids

$Co(CN)_5^{-3}$, Pd^{+2}, Pt^{+2}, Pt^{+4}

Cu^+, Ag^+, Au^+, Cd^{+2}, Hg^+, Hg^{+2}, CH_3Hg^+

BH_3, $Ga(CH_3)_3$, $GaCl_3$, $GaBr_3$, GaI_3, TI^+, $TI(CH_3)_3$

CH_2, carbenes

Pi-acceptors; trinitrobenzene, chloroanil, quinones, tetracyanoethylene, etc.

HO^+, RO^+, RS^+, RSe^+, Te^{+4}, RTe^+

Br_2, Br^+, I_2, I^+, ICN, etc.

O,Cl, Br, I, N , RO, RO_2.

M° (metal atoms) and bulk metals

Table 2.4 Classification of Hard and Soft Bases

Hard Bases

NH_3, RNH_2, N_2H_4
H_2O, OH^-, O^{-2}, ROH, RO^-, R_2O
CH_3COO^-, CO_3^{-2}, NO_3^-, PO_4^{-3}, SO_4^{-2}, ClO_4^-
$F^-(Cl^-)$

Borderline bases

$C_6H_5NH_2$, C_5H_5N, N_3^-, N_2
NO_2^-, SO_3^-
Br^-

Soft bases

H^-
R^-, C_2H_4, C_6H_6, CN^-, RNC, CO
SCN^-, R_3P, $(RO)_3P$, R_3As
R_2S, RSH, RS^-, $S_2O_3^{-2}$
I^-

Table 2.5 Bases in Order of Decreasing Hardness

Base	*Base*
1. H_2O	12. I^-
2. OH^-, OCH_8^-, F^-	13. SCN^-
3. Cl^-	14. SO_3^{-2}
4. NH_3	15. $(C_6H_5)_3Sb$
5. C_6H_5N	16. $(C_6H_5)_3As$
6. NO_3^-	17. $SeCN^-$
7. N_3^-	18. $C_6H_5S^-$
8. NH_2OH	19. $S{=}C(NH_2)_2$
9. $H_2N{-}NH_2$	20. $S_2O_3^{2-}$
10. C_6H_5SH	21. $(C_6H_5)_3P$
11. Br^-	

Pearson's Principle is "Hard acids prefer to bind to hard bases and soft acids prefer to bind to soft bases". Thus, then complex A:B is the most stable when both A and B are either hard or soft.

Pearson's principle provides the basis for an approximate qualitative prediction of the relative stability of the complexes, and is not an explanation of the observations.

The two simple theories which have been used to explain the preference of hard acids for hard bases and soft acids for soft bases are as follows:

(1) *Ionic and covalent bonding*: According to this concept, hard complexes involve ionic bonding and soft complexes result primarily from covalent bonding. It is expected that small sized and highly charged positive metal ions should favour ionic bonding with hard bases of class (a). When hard-soft combination takes place, the resulting species becomes unstable.

According to Misono and coworkers (1967) hardness and softness are correlated through the equation:

$$pK = -\log K = \alpha X + \beta Y + \gamma$$

where K is the instability constant of the complex, X and Y are parameters of the metal ion and α and β those of the ligand. γ is for adjustment so that all pK values lie on the same scale. For hard acids, the value of Y is found to be < 2.8 and for soft acids > 3.2. For borderline acids the value ranges between 2.8 and 3.2. The value of ligand parameter increases as we move from a hard base to a soft base.

(2) *Pi-bonding theory*: A characteristic of soft acids is the presence of loosely held outer d-electrons capable of being donated to ligands. Thus, most soft bases are π-bond acceptors and most soft acids are π-bond donors. Soft base include ligands containing P, As or I donor atoms and unsaturated ligands such as CO. Hard acids, on the other hand, are π-bond acceptors and hard bases (N, P, F) are π-bond donors. Again, the hard and soft interactions suffer from a mismatch in their bonding tendencies.

Applications of the HSAB Principle

Thermodynamic stability of simple compounds and complex ions can be rationalised in terms of hard and soft acid base principle.

AgI_2^- is stable but AgF_2^- does not exist. The reverse is true for cobalt (III) ; COF_6^{3-} is more stable than COI_6^{3-}. Table 2.3 shows that Ag^+ is a soft acid and Co (III) a hard acid; F^- is a hard and I^- a soft base. Ag^+ being a soft acid and I^- a soft base; both combine to give stable $[AgI_2]^-$. Similarly, Co^{3+} acid and F^- base (both being hard) combine to form stable $[CoF_6]^{3-}$.

With the help of this principle we can explain why $Hg(OH)_2$ dissolves readily in acidic aqueous solution but HgS does not. Mercuric sulphide may by regarded as a complex formed by combination of Hg^{2+}, a soft acid, and S^{2-}, a soft base whereas mercuric hydroxide is formed by the combination of Hg^{2+}, a soft acid with OH^-, a hard base. Hence HgS (soft acid + soft base) will be more stable than $Hg(OH)_2$ (soft acid + hard base).

The existence of certain ores can also be rationalised using the same principle. Hard acids such as Mg^{2+}, Ca^{2+} and Al^{3+} occur in nature as $MgCO_3$, $CaCO_3$ and Al_2O_3 respectively. This is because the cations and anions ($CO_3^{2-}$and O^{2-}) all are hard. These hard acid cations are never found in nature as sulphides since S^{2-} is a soft base. On the other hand, the soft acids such as Cu^+, Ag^+ and Hg^{2+} occur as sulphides. The borderline acids, Ni^{2+}, Cu^{2+} and Pb^{2+} occur in nature both as carbonates and sulphides.

The relative strength of halogen acids in aqueous solution can be predicted. In the reaction:

$$HX + H_2O \rightarrow H_3O^+ + X^- \quad \text{(where X = F, Cl, Br or I)}$$

the hard acid H^+ combines with the hardest base F^- to form the least dissociated or weakest acid, HF. Thus, the strengths of the halogen acids is in the order:

$$HF < HCl < HBr < HI$$

Hard solvents tend to dissolve hard solution and soft solvents tend to dissolve soft solutes.

The HSAB principle has been used to explain the poisoning of metal catalysts. Certain soft metals such as platinum and palladium act as catalysts in certain reactions. Bases containing phosphorus, arsenic, antimony, tellurium and selenium in low oxidation states (soft bases) poison these metallic catalysts, because they are strongly adsorbed on the surface and thus block the active sites. Thus soft metal catalysts are poisoned by impurities of soft bases, but remain unaffected by those of hard bases such as nitrogen, oxygen and fluorine.

The HSAB principle can be used to correlate the rates of certain chemical reactions. The speed of electrophilic and nucleophilic substitution reactions can be related to the hardness or softness of the acid and base centers. The reaction

$$H^+ + CH_3HgOH \rightarrow H_2O + CH_3Hg^+$$

should proceed toward the right hand side since it involves combination of hard acid H^+ and hard base OH^-. On the other hand the reaction

$$H^+ + CH_3HgSH \rightarrow H_2S + CH_3Hg^+$$

should proceed towards the left hand side as it involves combination of soft acid, CH_3Hg^+ and soft base S^{2-}.

SOME SOLVED EXAMPLES

Example 1. Which is the weakest Bronsted acid in water —HF, HCl, HI, or HBr?

Solution The reaction of the acid with water can be expressed (where $B = F^-$, Cl, Br^-, and I^-) as: $HB + H_2O \rightleftharpoons H_3O^+ + B^-$. Because H^+ is a hard acid and H_2O is a hard base, the hardest base, *B*, should be the most successful in competing with the water for the H^+ and should form the least dissociated or weakest Bronsted acid. F^- is the hardest base and, in fact, HF is the weakest of the acids.

Example 2. Which salt ionizes most in water, NaBr or $HgBr_2$?

Solution Br^- is a softer base than H_2O. Therefore the soft acid Hg^{+2} will prefer the combination $HgBr_2$ over $Hg(H_2O)_y^{+2}$ and will not ionize a great deal. The hard acid, Na^+, on the other hand, will prefer the hard base, H_2O, and will ionize extensively to form the $Na(H_2O)_2^+$ adduct. Another way to look at this effect is to think of polar water surrounding the dissociated ions, making their electrical charges. This hinders the association of their electrostatic attraction, but not the reassociation of those using primarily covalent bonding.

Example 3. Which complex is more stable in water, $Cd(CN)_4^{-2}$ or $Cd(NH_3)_4^{+2}$?

Solution The soft acid Cd^{+2} must compete for the hard base, H_2O, or the softer bases, NH_3 and CN^-. According to the HSAB rule, it will prefer the softer bases and both complexes will be stable in water. Of the two, however, CN^- is the softer and the cyano complex should be more stable. The $K_{\text{instability}}$ for $Cd(CN)_4^{-2}$ is 1.4×10^{-19} compared to that of $Cd(NH_3)_4^{+2}$ which is 7.5×10^{-8}.

Example 4. Which direction will the following reactions favour?

(*a*) $H^+ + CH_3HgOH \rightleftharpoons H_2O + CH_3Hg^+$

(*b*) $H^+ + CH_3HgS^- \rightleftharpoons HS^- + CH_3Hg^+$

Solution Reaction (*a*) goes to the right favoring the combination of the hard acid H^+ and the hard base OH^-. Reaction (*b*) goes to the left favoring the combination of the soft acid CH_3Hg^+ and the soft base S^{-2}.

Example 5. Explain why, in water, the acid H^+ appears to give the base strengths $OH^- > NH_3$, whereas Ag^+ gives the order $NH_3 > OH^-$.

Solution Again, the hard acid, H^+ prefers the harder base, OH^- and the soft acid, Ag^+, prefers the softer base, NH_3.

Example 6. Predict which way the following reactions will go:

(*a*) $CsI + LiF \rightleftharpoons LiI + CsF$

(*b*) $Ag(H_2O)_x^+ + HCl \rightleftharpoons AgCl + H_3O^+ + (x - 1)\,H_2O$

(*c*) $BeF_2 + HgI_2 \rightleftharpoons HgF_2 + BeI_2$

(*d*) $Co(CN)_5(H_2O)^{-3} + CN^- \rightleftharpoons Co\ (CN)_3^{-4} + H_2O$

(*e*) $Cd(H_2O)_6^{+2} + HS^- \rightleftharpoons CdS + H_3O^+ + 5H_2O$

Solution

(*a*) To the left—Cs^+ is softer than Li^+ (based on size). I^- is soft, F^- is hard.

(*b*) To the right—Cl^- is softer than H_2O; Ag^+ is soft, H^+ is hard.

(*c*) To the left—Be^{+2} is hard, Hg^{+2} is soft; F^- is hard, I^- is soft.

(*d*) To the right—$Co(CN)_5^{-3}$ is soft, CN^- is soft, and H_2O is hard.

(*e*) To the right—Cd^{+2} is soft, H^+ is hard; S^{-2} is softer than H_2O.

Example 7. Would you expect Pt^{+4} to form an S– or an N–bonded complex with SCN^-?

Solution An S-coordinated complex, Pt^{+4} is soft, S terminal is softer than N terminal.

Example 8. Which of these species is likely to be more stable? $[Co(CN)_5OH]^{-4}$ or $[Co(CN)_5SCN]^{-4}$

Solution $[Co(CN)_5SCN\ ]^{-4}$ is more stable. $Co(CN)_5^{-3}$ is soft, SCN^- is soft at S, OH^- is hard.

Example 9. Would you expect a metal ion M^{+n} to have the same hard or soft behavior in the gas phase as in an aqueous solution?

Solution No. The reacting species in the gas phase is M^{+n} whereas in the aqueous phase it is $M(H_2O)_x^{+n}$. In the aqueous phase, hard have to compete with the hard base H_2O for any hard acids. In addition, the polar water tends to mask the electrostatic forces characteristic of hard-hard interactions. Soft-soft interactions do not experience these problems to as great an extent. As a result, the contrast between hard and soft behaviour is much more significant in the aqueous phase than in the gas phase.

3

Solute–Solvent Interactions

THE ROLE OF THE SOLVENT

Most of the reactions carried out in researches and in industrial processes and a vast majority of those observed in nature occur in solution. The molecules, atoms, or ions of the various reactants are often completely or partially dispersed in some, usually liquid but sometimes solid or gaseous, substance which serves as the medium for the reaction. The effects of the solvent on the course of reactions have been frequently ignored. Up to the turn of the century chemists were concerned chiefly with reactions in which water served as the solvent medium.

It is, in fact, true that the influence of the solvent on the course of a chemical reaction can be profound, indeed. By changing the solvent, the products from a given set of reactants can be completely changed, and in some cases, reactions may be reversed. For example, in aqueous solution silver nitrate and barium chloride react to form a precipitate of silver chloride leaving barium nitrate in solution, whereas in liquid ammonia solution silver chloride combines with barium nitrate to yield a precipitate of barium chloride, leaving silver nitrate in solution.

ENERGETICS OF SOLVATION

The chemistry of ionizing solvents is essentially the study of the properties of those solvents and their solutions in which the interaction with solutes, be they covalent compounds or ionic crystals, is sufficiently strong to produce ions in solutions. In view of the strength of the forces holding ionic crystals together and the size of the energies required to produce heterolytic fission of covalent bonds the energies of these solvent-solute interactions must be very considerable, probably between 200 and 400 kJ mol^{-1} per g-ion produced.

It must be admitted at this stage that there is no adequate quantitative theory which will predict the nature and size of these forces for a wide range of solvents and that even a completely adequate qualitative theory is lacking, although a number of useful empirical guidelines to behaviour will be discussed later in this chapter. However, once the ions are produced in solution, then at moderate dilution it is possible to rationalize their behaviour and conductivity by theories in which the only solvent parameter is the dielectric constant.

The earliest quantitative treatment of the solvent-ionic solute interaction attempted to rationalize the energy of interaction of an ion with its solvent in terms of the behaviour of a dielectric in the field of a charged sphere. Simple electrostatic arguments show that if a sphere of radius R, carrying a charge q, is moved from vacuum to a solvent of dielectric constant ε, there is a free energy change:

$$\Delta G = \frac{q^2}{2R}\left(1 - \frac{1}{\varepsilon}\right)$$

or, per g-ion:

$$\Delta G = -\frac{N_A z^2 e^2}{2R}\left(1 - \frac{1}{\varepsilon}\right)$$

where ze is the charge on the ion, e being the electronic charge and z the valency, and N_A is Avogadro's constant. The corresponding enthalpy change is;

$$\Delta H = -\frac{N_A z^2 e^2}{2R}\left(1 - \frac{1}{\varepsilon} - \frac{T}{\varepsilon^2}\cdot\frac{\partial_\varepsilon}{\partial T}\right)$$

Such a theory is clearly a gross approximation, for the solvent near the ion cannot be treated as a continuous dielectric and in the ion's immediate neighbourhood the electric fields are so intense that the solvent molecules are likely to be coordinated. In addition there is the problem of what value one takes for the radius of the ion in solution. However, the formulae do offer an estimate of the energies involved and in Table 3.1 values are given of the energies of solvation on this model of different values of R (in Å) and ε.

Some general points emerge from this table. For really small ions like Li^+ (crystal radius = 0.68) and Na^+ (crystal radius = 0.98 Å), the solvation energies are very large and it is not surprising to find that salts containing these cations, particularly when the anions are large

Table 3.1 Effect of Dielectric Constant on Free Energy of Solvation

R \ ε	2	4	8	16	32	64	128	∞
0.5	694	1041	1216	1300	1346	1362	1375	1387
1.0	347	518	606	652	673	681	690	694
1.5	230	347	405	434	447	455	460	464
2.0	175	260	305	326	334	342	342	346
3.0	117	175	201	217	225	225	230	230
4.0	88	129	150	163	167	171	171	171
5.0	71	104	121	130	134	134	137	137

The 'Born charging equation' is used, R is the radius of the ion in Å, ε the dielectric constant of the solvent. Free energies are given in kJ mol^{-1}.

and the lattice energies consequently low, are soluble in solvents of quite low dielectric constant. Second, the change in the solvation energy on increase of dielectric constant above a value of ε of about 30 is relatively small, and so one would expect all materials with dielectric constants above 30 to be quite good ionizing solvents. Table 3.2 lists the dielectric constants of some ionizing solvents.

Table 3.2 Dielectric Constants of a Number of Ionizing Solvents at 25°C

Solvent	ε	*Solvent*	ε
H_2SO_4	100	I_2	11.1 (118°C)
$CH_3.CO.N(CH_3)_2$	37.8	Br_2	3.12 (20°C)
$CH_3.CO.NH(CH_3)$	165.5 (40°C)	$AsCl_3$	12.6 (17°C)
$(CH_3)_2SO$	46.6	$SbCl_3$	33.0 (75°C)
$CH_3.NO_2$	35.9	SO_2	15.4 (0°C)
$C_6H_5.NO_2$	34.8	NH_3	33 (−33°C)
$(CH_3)_2CO$	20.7	16.90	
$CH_3.OH$	32.6	HF	175 (−73°C)
$C_2H_5.OH$	24.3	84 (0°C)	
C_5H_5N	12.3	HCl	9.28 (−95°C)
$CH_3.CN$	36.2	HBr	7.0 (−85°C)
NOCl	22.5 (−27.5°C)	H_2O	78.30
$POCl_3$	13.9 (20°C)		

Since liquids with a dielectric constant of over 30 are inevitably associated to some degree, other physical properties tend to be related to good solvent behaviour. Association usually means that the freezing and boiling points are higher than one would expect from a simple consideration of the molecular weight, e.g. compare the boiling points of water, ammonia, and hydrogen fluoride with those of hydrogen sulphide, phosphine, and hydrogen chloride. One would also expect that, as associated liquids, good solvents would exhibit anomalous Trouton's constants with higher values than that of 21.5 found for unassociated liquids.

The Nature of Solvent-Solute Interactions

Modifications of the continuum theory of ion solvation outlined above have usually considered that the solvent molecules in the first layer around the ion in solution have their dipoles oriented by the ionic charge and are coordinated either by ion-dipole forces or by coordinate bonds. Interactions outside the first coordination sphere are then treated by a continuum model. This type of treatment has produced, in the case of water, good agreement between the sums of theoretically calculated solvation energies for pairs of positive and negative univalent ions and the experimental values. Clearly a high solvent dipole moment will favour a high interaction energy with a solute ion. Some solvent dipole moments are given in Table 3.3.

Table 3.3 Dipole Moments of a Number of Ionizing Solvents

Solvent	*Dipole moment (Debye units)*	*Solvent*	*Dipole moment (Debye units)*
H_2O	1.85	$CH_3.OH$	1.70
HF	1.82	$C_2H_5.OH$	1.69
NH_3	1.47	$AsCl_3$	1.59
HCl	1.08	$SOCl_2$	1.45
$(CH_3)_2SO$	3.96	NOCl	1.83
C_5H_5N	2.1	NO_2	0.316
$CH_3.CN$	3.92	$(CH_3)_2CO$	2.88
SO_2	1.63	$CH_3.NO_2$	3.46
$CH_3.CO.N(CH_3)_2$	3.81	$C_6H_5.NO_2$	4.22
$CH_3.CO.NH(CH_3)$	3.73		

Ion–dipole interactions are obviously the main source of the solvation energy of the alkali metal ions, but when transition metal ions are considered actual donation of an electron pair *from* solvent molecules to the ion takes place and, clearly, physical measurements that assess the power of coordination will give a measure of solvent power. With negative ions, electrostatic ion–dipole forces are for many solvents the only type of interaction that can take place. There are, however, two fairly specific types of interaction that can occur in some systems between the solvent and negative ions.

Hydrogen bonds may form when the solvent possesses polar hydrogen atoms. Another type of interaction that may occur is donation of electron pairs from the negative ion to the solvent.

The strongest and most effective solvents are those which, in addition to having a high dielectric constant and a largish molecular dipole moment, can interact with both positive and negative ions in one of the specific ways mentioned above. Thus, water forms strong coordinate bonds with transition metal ions and can also hydrogen bond strongly with negative ions such as fluoride, chloride, bromide, hydroxide, etc.

In some cases, usually when transition metal ions and halide ions are involved, there is an actual competition between the halide ions and the solvent in coordination with the metal ion. Consider, for example, cobaltous chloride, $CoCl_2$. In dilute aqueous solutions, the ions present the $Co(OH_2)_6^{2+}$ and Cl^-, and the solution has a characteristically pale pink colour. In solvent of lower dielectric constant and of lower coordination power—such as acetone—a whole range of species $CoCl_4^{2-}$, $CoCl_3S^-$, $CoCl_2S_4$, $CoClS_5^+$, and CoS_6^{2+} occur. Such solutions are characterized by a deep blue colour due to the tetrahedral species $CoCl_4^{2-}$ and $CoCl_3S^-$ (S represents a solvent molecule).

Measurements of 'Solvent Strength'

A wide variety of physical techniques has been suggested for use as a measure of solvent donor strengths, though relatively few as a measure of solvent acceptor strengths, and some measurements have been used as a measure of 'solvent strength' in general.

In the vast majority of solvents an ion such as Ni^{2+} is octahedrally coordinated by solvent molecules. Ligand field theory tells us that in such a situation the energy levels of the *d*-electron orbitals on the nickel ion will be split into two groups, the e_g and t_{2g} levels.

Transitions between these levels are possible and they give rise to the characteristic ligand-field absorptions in the electronic spectrum of the coordinated ion. Such transitions occur in the visible and near-infrared regions of the electromagnetic spectrum. Measurement of their wavelengths enables us to calculate the energy different between the e_g and the t_{2g} levels. This difference is usually denoted by 10 Dq and the value of Dq in cm^{-1} gives a measure of the solvent donor strength.

A property which should correlate better with the dielectric constant is the equilibrium constant for the association into ion pairs $K_{assoc.}$, of the tetra-alkylammonium salts. The attractive force between two ions of opposite charge, ze, separated by a distance r in a homogeneous medium of dielectric constant ε is given by:

$$F = \frac{z^2 e^2}{\varepsilon r^2}$$

In a solvent two effects will occur:

1. The charge-dipole interactions of the ions with the solvent will result in a partial charge distribution over the first layer of solvent molecules round the ion. This dispersion of charge will attenuate the electric field round the ion and decrease ion pairing.

2. A high bulk dielectric constant will attenuate the electric field around the ion more rapidly than a low dielectric constant and again reduce ion pairing.

Some values of $K_{assoc.}$ for tetra-alkylammonium salts in different solvents are given in Table 3.4. For solvents with dielectric constants much above 30 the values of $K_{assoc.}$ cease to be a useful guide, but there is a wide spread of values for those solvents with values of ε less than 30.

Another criterion for the coordinating power of a solvent, suggested by Joesten and Drago, is the heat of formation of the phenol adduct. This measures fairly specifically the ability of the solvent to act as an electron pair donor or hydrogen bond formation.

An alternative criterion of coordinating power is the heat of formation of the adduct with iodine. Here one is measuring the ability of the solvent to act as an electron pair donor to a very 'soft' Lewis acid.

Table 3.4 **Values of $K_{assoc.}$ for a Variety of Solvents**

Solvent	*Salt*	$K_{assoc.}$
$CH_3.CO.N(CH_3)_2$	Et_4NBr	20
	Pr_4NBr	20
$CH_3.CO.NH(CH_3)$	Et_4NBr	0
	CsBr	0
$C_6H_5.NO_2$	Bu_4NBr	46
$(CH_3)_2CO$	Bu_4NI	164
$CH_3.OH$	Bu_4NBr	26.0
C_5H_5N	Bu_4NBr	4,000
	Bu_4NI	2,440
$CH_3.CN$	Me_4NCl	77.5
	Me_4NBr	41.4
	Me_4NI	27.5
SO_2	Me_4NCl	10,300
	Me_4NBr	11,800
	Me_4NI	13.900
	Et_4NBr	21,000

EFFECT OF PHYSICAL PROPERTIES OF THE SOLVENT

The range of applicability of various solvents to the carrying out of chemical reactions is determined by the physical properties of the solvent.

Melting Point and Boiling Point

Since most reactions in solvent systems are most conveniently carried out in the liquid phase, the temperature range between the melting point and boiling point of the solvent virtually establishes the temperature range of usefulness of the solvent as far as reactions at atmospheric pressure are concerned. This range can be extended by use of diluent to lower the freezing point and raise the boiling point of the solvent or by operating at increased pressure to raise the boiling point. However, such measures are sufficiently inconvenient to militate considerably against the use of a particular solvent unless some other aspect of the process demands that particular solvent. In the case of liquid ammonia, the abundance and low cost of the solvent combined with its interesting and unusual solvent characteristics cause it to be widely used even though its liquid range at atmospheric pressure (–77.7°C to –33.35°C) is not particularly desirable.

Heat of Fusion and Heat of Vaporization

Since the intermolecular forces in the solid state must be largely overcome when a solid melts, and since intermolecular forces in the liquid phase oppose the vaporization of the liquid it is clear that the magnitudes of the molal heats of fusion and vaporization give us some insight into the nature and strength of the associative forces between molecules in these condensed phases. The degree of chemical association of molecules in a liquid determines to no small extent its solvent properties. Therefore, the heat of fusion and, particularly, the heat of vaporization are significant in evaluating the utility of a solvent.

The ratio of the heat of vaporization to the boiling point on the absolute scale (°K) is a constant called the Trouton constant for many (so-called "normal") liquid, and for these liquids it has the value 21.5. A higher value indicates association of molecules of the liquid to form larger aggregates. Such abnormal liquids include some of our most useful solvents, e.g., water, liquid ammonia, liquid hydrogen fluoride, and the alcohols. One of the most common factors which result in molecular association in the liquid state is polarity of the molecule. Polarity of a molecule results when unsymmetrical molecules contain bonds in which the electron pair forming the bond is shared unequally between the two bonded atoms. Another way of stating this is that polar molecules are molecules in which the center of negative charge (electronic charge) and the center of positive charge (nuclear charge) do not coincide. This is illustrated by the following structures for iodine monochloride and ethyl alcohol.

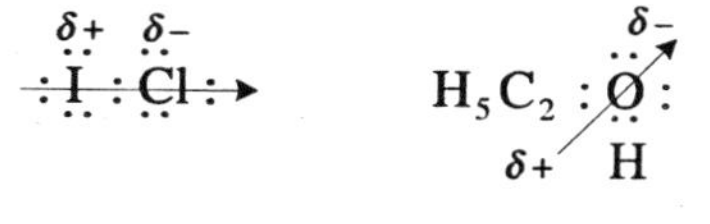

$$\overset{\delta+}{:\ddot{\underset{..}{I}}}:\overset{\delta-}{\ddot{\underset{..}{Cl}}}: \longrightarrow \qquad\qquad H_5C_2:\overset{\delta-}{\ddot{\underset{..}{O}}}: \quad \underset{\delta+}{} H$$

Association between polar molecules arises from the tendency of polar molecules to orient themselves so that the positive ends of the molecules are near the negative ends of the neighboring molecules.

One effect of the polarity of a solvent is that polar solutes tend to be much more soluble in liquids of high polarity than they are in nonpolar liquids. Where the solute is polar it can enter into association with the molecules of the polar solvents and thus more readily become dispersed in the solvent than would be the case where solvent and solute are strongly different in polarity. Thus, gasoline, composed of virtually non-polar hydrocarbon molecules, is almost completely insoluble in the highly polar solvent water, but is completely miscible with the non-polar solvent carbon tetrachloride. On the other hand, the polar substance glucose is highly soluble in water but is virtually insoluble in carbon tetrachloride.

Dielectric Constant

According to Coulomb's well-known law, the force between two charged bodies is given by the expression

$$F = \frac{e_1 e_2}{D r^2}$$

where e_1 and e_2 are the electrical charges on the two bodies, r is the distance between the two charges, and D is a constant called the *dielectric constant* which depends upon the nature of the medium in which the two charged bodies are suspended. It is clear that if D is large the force F will be small. In such a case the medium is said to be an insulating medium. D is usually assigned a value of 1 for a vacuum. Liquids made up of polar molecules (e.g., water and ammonia) usually have very high dielectric constants; for example, the dielectric constant of water at room temperature is about 80.

In order for an ionic solid to dissolve, the electrostatic forces which stabilize the crystal must be greatly reduced. It is clear that this may be accomplished if the ions become suspended in a medium of high dielectric constant. It is, therefore, to be expected that, in general, solvents of high dielectric constant will be much better solvents for ionic substances than are solvents of low dielectric constant. Sodium chloride, an ionic, crystalline substance, dissolves readily in water but is virtually insoluble in the non-polar liquid carbon tetrachloride.

Viscosity

One of the striking characteristics of various liquids is the differences in their rates of flow. Certain ones such as water, ethyl alcohol, and carbon tetrachloride, are at ordinary temperatures highly fluid and flow rapidly under gravitational force; these are said to be highly mobile. Others such as certain high-molecular weight hydrocarbons and anhydrous sulfuric acid are highly viscous and flow at much lower rates under a given set of conditions.

Factors affecting viscosity are the temperature of the liquid and the size and shape of the molecule. Generally lowering the temperature, increasing the molecular size, or making the molecule less symmetrical increases the viscosity of a liquid.

Increasing the viscosity of a liquid lowers the mobility of ions and molecules in the liquid thus reducing the electrical conductance of solutions of electrolytes in the liquid, and greatly increasing the difficulty of such operations as precipitation, crystallization, and filtration involving that particular liquid.

EFFECT OF CHEMICAL PROPERTIES OF THE SOLVENT

Of great significance in the selection of solvents for particular chemical applications are the limitations imposed by the chemical nature of the solvent. A reagent used in a given solvent must be one toward which the solvent is either inert, or, at the very least, much less reactive than the substance with which it is supposed to react. Stated in another way, the only reagents available for reaction in a given solvent are those capable of existing for an appropriate period of time in that solvent.

Effect of Acidic or Basic Characteristics of the Solvent

The acidity or basicity of a protonic solvent exerts a most profound effect on the usefulness of the solvent because of what is known as the leveling effect of the solvent. This effect results from the fact that the strongest acid which is available in a given protonic solvent is that which results from the self-ionization off the solvent. Likewise, the strongest base available is that which results from the self-ionization of the solvent. Consider, for example, the following auto-ionization reactions:

$$2H_2O \rightleftharpoons H_3O^+ + OH^-$$
$$2NH_3 \rightleftharpoons NH_4^+ + NH_2^-$$
$$2CH_3COOH \rightleftharpoons CH_3COOH_2^+ + CH_3COO^-$$

In aqueous solution the strongest acid which can exist and which, therefore, can be available for reaction is the hydronium ion H_3O^+. All the so-called strong acids familiar to the chemist appear to have exactly equal strength in aqueous solution. This results from the fact that these acids, such as perchloric ($HClO_4$), hydriodic (HI), hydrobromic (HBr), nitric (HNO_3), and hydrochloric (HCl) all are stronger proton donors than is the hydronium ion and, hence, all react completely with water to produce hydronium ion.

$$\mathrm{H : X + H : \ddot{\underset{\cdot\cdot}{O}} : \underset{H}{} \rightarrow H : \ddot{\underset{\cdot\cdot}{O}} : H^+ + : X^-}$$
$$\text{(with H below each O)}$$

Thus, acids stronger than H_3O^+ are leveled to the strength of H_3O^+ when placed in aqueous solution.

Similarly, the very strong bases such as hydride ion (H^-), amide ion (NH_2^-), and ethoxide ion ($OC_2H_5^-$), all react completely in aqueous solution to yield hydroxide ion (OH^-).

$$\mathrm{X^- : + H : \underset{H}{\ddot{\underset{\cdot\cdot}{O}}} : \rightarrow H : X + : \ddot{\underset{\cdot\cdot}{O}} : H^-}$$

It is, therefore, fruitless to use such strong bases in aqueous solution for they are immediately leveled to the base strength of hydroxide ion through reaction with the solvent. For example, when sodium hydride, NaH, which contains the very strong base hydride ion (H^-), is added to water a vigorous reaction ensues with the liberation of hydrogen

$$\mathrm{Na^+,\ : H^- + H : \underset{H}{\ddot{\underset{\cdot\cdot}{O}}} : \rightarrow Na^+,\ : \ddot{\underset{\cdot\cdot}{O}} : H^- + H : H}$$

and the solution is found to be a solution of sodium hydroxide. Similar reactions ensure when other bases stronger than hydroxide ion are added to water. The reactions with sodium amide and sodium ethoxide are further examples

$$\mathrm{Na^+,\ : \underset{H}{\ddot{\underset{\cdot\cdot}{N}}} : H^- + H : \underset{H}{\ddot{\underset{\cdot\cdot}{O}}} : \rightarrow Na^+,\ : \ddot{\underset{\cdot\cdot}{O}} : H^- + H : \underset{H}{\ddot{\underset{\cdot\cdot}{N}}} : H}$$

$$\mathrm{Na^+,\ : \ddot{\underset{\cdot\cdot}{O}} : C_2H_5^- + H : \underset{H}{\ddot{\underset{\cdot\cdot}{O}}} : \rightarrow Na+,\ : \ddot{\underset{\cdot\cdot}{O}} : H^- + : \underset{H}{\ddot{\underset{\cdot\cdot}{O}}} : C_2H_5}$$

It follows, therefore, that the intrinsic acidity and basicity of protonic solvents have a profound effect in determining the usefulness of the solvent is reactions where the availability of highly acidic or basic reagents is important. In the chapters which follow we shall for the various solvents discussed consider examples of such effects.

In terms of the above discussion it is clear that if we wish a solvent in which the intrinsic differences in the proton-donor tendencies of various acids will be evident, we will choose a solvent which has relatively weak proton-acceptor tendencies. Thus, as far as acids are concerned, liquid ammonia is a *leveling* solvent, since all strong acids are leveled to the acidity of the weak acid ammonium ion, NH_4^+. On the other hand, anhydrous acetic acid, being a relatively poor proton acceptor, acts as a *differentiating* solvent for acids since the intrinsic differences in acidity of such acids as perchloric, and hydrochloric become evident in this solvent. Perchloric acid undergoes virtually complete reaction with the acetic acid solvent

$$CH_3C(=O)—OH + HClO_4 \rightarrow CHC(OH)_2^+ + ClO_4^-$$

but sulphuric and hydrochloric acids behave as weak electrolytes in this solvent.

In an analogous manner acetic acid is a leveling solvent toward bases, for anhydrous acetic acid reacts completely with most of the common bases such as CN^-, OH^-, $OC_2H_5^-$, and NH_2^-, leveling all of them to CH_3COO^-. Strongly basic solvents such as liquid ammonia, however, act as differentiating solvents toward bases.

Aprotic solvents, which have neither strong proton-donor tendencies nor strong proton-acceptor tendencies, may act as differentiating solvents for both acids and bases. In such instances the solvent serves principally as the suspending medium for the solute species and itself participates in chemical reactions in the solvent to only a minor extent.

Chemical Effects on Solubility

In addition to the effects already described it should be noted that the solubility of a specific solute in a specific solvent can be greatly increased by chemical interactions between solute and solvent species. For example, acetone is a solvent of very low dielectric constant and the acetone molecule is of low polarity. Yet such protonic solvents as water and ethyl alcohol are miscible with acetone in all proportions. The principal reason for these high solubilities of water and ethyl alcohol in acetone is the interaction of molecules of these compounds with acetone through the mechanism of hydrogen bonding.

$$H-O-H \cdots O=C(CH_3)_2 \qquad C_2H_5-O-H \cdots O=C(CH_3)_2$$

Other examples of this effect include the high solubilities of carbon dioxide and sulfur dioxide in water which result, in part, from their reactions with water to form carbonic and sulfurous acids, respectively.

$$CO_2 + H_2O \rightleftharpoons H_2CO_3$$
$$SO_2 + H_2O \rightleftharpoons H_2SO_3$$

EFFECTS OF OXIDIZING AND REDUCING CHARACTERISTICS

The chemistry in a given solvent can be strongly affected by the acidity or basicity of the solvent as well as by chemical effects on solubilities in the solvent. This influence of solvent on the nature of chemical processes which may be carried out in solution in it also extends to the field of oxidation-reduction processes. Water provides an excellent example, for we find that aqueous chemistry does not make available strong reducing agents. The reason for this is that water is so susceptible to reduction (with the release of hydrogen) that strong reducing agents react immediately with water. Hence, only those reducing agents whose electrode potentials are below hydrogen in the electrode potential series are available for aqueous reactions. Another example is provided by liquid ammonia in which strong oxidizing agents may not be used because ammonia is itself a reducing agent and strong oxidizing agents oxidize the ammonia molecules to elementary nitrogen or other oxidation products.

At this point it is discussing, in general, acid-base reactions in a solvent and self-ionization of solvents. It must be made clear that ionizing solvents, that is, solvents capable of producing ionization in dissolved solutes need not themselves ionize to any appreciable extent. Indeed the self-ionization of most organic solvents is negligible. However, there are two classes of solvent which do exhibit appreciable self-ionization. The first class comprises solvents with very polar bonds to a hydrogen atom—usually the more polar the bond the greater the degree of self-ionization. Thus, ammonia would be expected to exhibit a lower self-ionization than water. The most highly self-ionized solvents are the Group VI acids, sulphuric acid and selenic acid. Phosphoric acid also exhibits a very high degree of self-ionization.

Table 3.5 Auto-Solvolysis Constants, K_{ap}, and Specific Conductivities of Pure Solvents, L_0

Solvent	$-\log_{10} K_{ap}$	L_0 (ohm^{-1} cm^{-1})
NH_3	29.8	1×10^{-11}
$C_2H_5.OH$	18.9	
H_2O	14.0	
$CH_3.CO_2H$	12.6	
HF	9.7	$<10^{-6}$
H_2SO_4	3.6	1.0439×10^{-2}
H_3PO_4	~2	
$AsCl_3$	<15	1.4×10^{-7}
$POCl_3$	<13	2×10^{-8}
BrF_3		8.0×10^{-3}
ICl		4.6×10^{-3}

The best quantity to use to compare self-ionizations is the autosolvolysis constant of the solvent. Unfortunately the information necessary to obtain this is not always available and one is often reduced to comparing specific conductivities of the pure solvents. A list of these for various solvents is given in Table 3.5.

The other class of solvents which shows appreciable self-ionization is very polar halides. Here the effect is often accompanied by three-centre bonding through a halogen atom bridge with a considerable increase in solvent viscosity. An example off this class is antimony pentafluoride.

Another general point that may be considered here is the possible range of acid-base strengths in a solvent. In terms of Bronsted-Lowry or protonic acids, the strongest acid in a solvent is the protonated form of that solvent and the strongest base the deprotonated form. Thus, the strongest acid in water is the H_3O^+ ion and the strongest base the OH^- ion. In liquid ammonia the NH_4^+ ion is the strongest acid and the NH_2^- ion the strongest base. The basicity of ammonia is about 10^{12} times greater than that of water and its acidity about 10^{-25} times less. Consequently, acids with a pK_a less than about 12 in water are 'strong' acids in liquid ammonia. Acetic acid is a strong acid in liquid ammonia, as is ammonium acetate.

Any solvent which accepts protons more readily than water will be a more basic solvent than water, and any solvent which donates protons to a solute more readily than water, i.e. deprotonates more readily, will be a more acid solvent than water. The smaller the self-ionization constant of the protonic solvent, the wider the range of acid strengths that may be studied in it.

When one moves away from considering protonic acids and bases it is no longer possible to be so quantitative as the necessary data are not available. However, the same principles hold. Consider the Lewis acid boron trifluoride. Its strength as an acid in a solvent is determined by the strength of its complex with the solvent. In pyridine, for example, the complex $C_6H_5N \cdot BF_3$ is formed and only Lewis bases stronger than pyridine will displace pyridine from the complex. Acetonitrile will not and so is not powerful enough to function as a Lewis base in pyridine.

Again, consider the reactions of the chloride ion, a Lewis base, with the weakish Lewis acid, germanium tetrachloride. In liquid hydrogen chloride, the chloride ion interacts strongly with the solvent, via hydrogen bonds, to form the bichloride ion, HCl^-_2. Consequently, the hexachlorogermanate ion is not formed. In a solvent, such as nitromethane, that interacts less strongly with chloride ions, $GeCl_6^{2-}$ ions are formed.

Just as the properties of solvent limit acid and base strengths to a range of values, so the properties of a solvent limit the range of oxidation potentials attainable in it and hence the range of oxidizing agents and reducing agents that can be used. Thus, in liquid ammonia, if all reactions were thermodynamically controlled (as opposed to being kinetically controlled) no oxidizing agent more powerful than nitrogen and no reducing agent more powerful than hydrogen could exist in the solvent. In acid solution the potentials of the NH_3–N_2 and H_2–NH_3 half-cell reactions are:

$$\tfrac{1}{2}H_2 + NH_3 = NH_4^+ + e^- \qquad E^0 = 0$$

$$4NH_3 = \tfrac{1}{2}N_2 + 3NH_4^+ + 3e^- \quad E^0 = 0.04$$

and the range of potentials in basic solution is the same. With only a range of 0.04 V available (compare 0.815 V for pure water and 1.23 V for aqueous acid solutions) hardly any oxidizing or reducing species would be stable in liquid ammonia. However, both the hydrogen couple and particularly the nitrogen couple show high overvoltages in the solvent, so that one can work with a range of thermodynamically unstable, but kinetically stable, oxidizing and reducing agents.

Table 3.6 Thermodynamic Limits of Oxidation-Reduction Reactions in a Range of Solvents

Solvent	*Highest oxidized species*	*Lowest reduced species*
NH_3	N_2	H_2
H_2O	O_2	H_2
HF	F_2	H_2
HCl	Cl_2	H_2
H_2SO_4	$H_2S_2O_8$	H_2
HSO_3F	$S_2O_6F_2$	H_2
$CH_3.OH$	H_2CO	H_2
$CH_3.COOH$	$(CH_3.CO)_2O_2$	H_2
$SbCl_3$	$SbCl_5$	Sb (?)
$AsCl_3$	Cl_2	As

Many solvents offer a much wider thermodynamic range than liquid ammonia. Thus, in liquid hydrogen fluoride any compound with an oxidation potential below that of fluorine or above that of hydrogen would be thermodynamically stable. The oxidized and reduced species which set the upper and lower limits of thermodynamic stability in a variety of solvents are listed in Table 3.6.

SOLVENT CHARACTERISTICS OF WATER

It is useful to summarize briefly the principal characteristics of the solvent water.

1. Water freezes at 0°C and boils at 100°C under one atmosphere pressure and, therefore, has a very convenient liquid temperature range.

2. Water is itself an exceedingly poor electrical conductor, undergoing autoionization according to the equation $2H_2O \rightleftharpoons H_3O^+ + OH^-$ only to an exceedingly small extent.

$$[H_3O^+][OH^-] = 1 \times 10^{-14} \text{ at } 25^\circ C$$

3. The dielectric constant of water at 20°C has the very large value 80 and water is, therefore, an excellent solvent for many electrovalent

compounds. The resulting solutions are excellent conductors of electricity.

4. Many covalent compounds react with water to give electrically conducting solutions, for the hydration of many covalent compounds results in the production of ions, e.g.

$$HCl + H_2O \rightarrow H_3O^+ + Cl^-$$
$$Al_2Cl_6 + 12H_2O \rightarrow 2Al(H_2O)_6^{+++} + 6Cl^-$$
$$NH_3 + H_2O \rightleftharpoons NH_4^+ + OH^-$$

5. Water is an excellent solvent in which to carry out neutralization reactions, for reactions of this type take place rapidly and smoothly in aqueous solution.

6. Water readily hydrates many salts, acids, and bases through the processes of coordinate covalent bond formation, ion-dipole attraction, or hydrogen-bonding.

7. Reactions involving acids stronger than H_3O^+ ion cannot be carried out in aqueous solution. Likewise, reactions involving bases stronger than OH^- ion may not be carried out in aqueous solution.

8. Reactions involving strong reducing agents may not be carried out in aqueous solution because such reagents would react with water, resulting in the evolution of hydrogen. Moderately strong oxidizing agents may be used in aqueous solution.

9. Because of the labile character of the hydrogen to oxygen bond, hydrolytic reactions proceed rapidly in aqueous solutions. Not only do many salts of weak bases or weak acids undergo hydrolysis but also many acid halides, esters and other covalent substances readily hydrolyze in contact with water.

These then are the characteristics which, along with its abundance, principally account for the widc utilization of water as a solvent. They also indicate important limitations in its usefulness.

Type Reactions in Solvents

Whereas it may readily be shown that virtually all chemical reactions are modified to a degree by the nature of the solvent in which they are carried out, there are certain type of reactions which are more influenced by the nature of the solvent than are others and should at this point, therefore, be briefly noted. Four types of reactions belonging to this category are: (1) precipitation, (2) salt formation, (3) solvolysis, and (4) solvation.

Precipitation

The formation of a precipitate when solutions of two compounds are mixed is one of the most familiar of all chemical processes. Such a manifestation is dependent upon the solubilities of the possible products in the particular solvent as well as upon the various other equilibria involving one or more of the ions composing the product and other components of the solution. Since the values of these solubilities depend in very instance on the nature of the solvent used, it should come as no surprise to find that such precipitation reactions also are highly dependent on the solvent employed. The example of the reversal of the following process

$$BaCl_2 + 2AgNO_3 \rightarrow 2AgCl + Ba(NO_3)_2$$

when the solvent is changed from water to liquid ammonia was given at the beginning of the chapter. Likewise, sodium hydroxide, highly soluble in water, can be precipitated in liquid ammonia; water-soluble ammonium sulfate is virtually insoluble in liquid ammonia, and water-soluble copper sulfate may be precipitated in anhydrous acetic acid.

Salt Formation

If one is seeking to prepare a certain salt, one must bring together an appropriate acid and base, sufficiently acidic and basic, respectively, to react virtually completely with each other. In such an instance, the solvent must be one in which the appropriate acid and base can exist.

Thus for example, if we wish to prepare the salt $Na^+[H_2N{-}\overset{\overset{O}{\|}}{C}{-}NH^-]$, we must use a base capable of taking a proton from the exceedingly weak acid urea. Since the strongest base available in water viz. the hydroxide ion, will not take a proton from the urea molecule, the salt $Na^+[H_2N{-}\overset{\overset{O}{\|}}{C}{-}NH^-]$ may not be prepared in aqueous solution. Viewed another way, the salt is unstable in water and the ureide ion will take a proton from water to form hydroxide ion.

$$Na^+\ H_2N{-}\overset{\overset{O}{\|}}{C}{-}NH^- + H_2O \rightarrow Na^+\ OH^- + H_2N{-}\overset{\overset{O}{\|}}{C}{-}NH_2$$

The sodium salt of urea may, however, be readily formed in liquid ammonia solution by the reaction of urea with the strong base sodium amide.

$$H_2N{-}\overset{\displaystyle /\!\!/O}{C}{-}NH_2 + Na^+\ NH_2^- \rightarrow Na^+\ H_2N{-}\overset{\displaystyle /\!\!/O}{C}{-}NH^- + NH_3$$

Another interesting example is provided by the salt nitronium perchlorate, $NO_2^+\ ClO_4^-$. This salt is formed by the reaction of the very strong acid perchloric acid, $HClO_4$, with the *exceedingly weak base* O_2NOH. (Note that this substance is the well-known strong acid, nitric acid.) This reaction may not be carried out in water, for nitronium ion is incapable of existence in aqueous solution.

$$NO_2^+ + 2H_2O \rightarrow NO_2OH + H_3O^+$$

In the strongly acid solvent anhydrous sulfuric acid, however, the reaction occurs readily.

$$NO_2OH + HClO_4 \xrightarrow{H_2SO_4} NO_2^+,\ ClO_4^- + H_2O$$

Solvolysis

A solvolytic reaction is a reaction in which the solvent molecule reacts with the solute in such a way that the solvent molecule is split into two parts, one or both of which become attached to a solute molecule or ion. In most instances the solvolytic process results in the development of an increased concentration of either the cation or anion which is characteristic of the autoionization of the solvent.

The terms *hydrolysis, ammonolysis, acetolysis,* and *alcoholysis* are applied to solvolytic reactions in which water, ammonia, acetic acid, and ethyl alcohol are, respectively, the solvents concerned.

Solvation

A solvation reaction is formally a reaction in which a molecule of the solvent attaches itself to a solute species (cation, anion, or molecule) by any of the various types of chemical bonds, notably ion-dipole, hydrogen bonding or coordinate covalent bonding. In certain instances solvation is difficult to distinguish from solvolysis, for the first steps in many solvolytic processes consist of solvation reactions. However, the term solvation is usually applied to those reactions where the solvent molecule remains in the resulting product, though in fact a bond in the solvent molecule may be broken.

Solvation of a cation occurs through ion-dipole interaction between the positive cation and the negative end of the solvent dipole or through coordinate covalent bonds in which the solvent molecule shares a pair of electrons with the cation. Solvation of a negative ion can occur through ion-dipole interactions between the negative anion and the positive end of the solvent dipole, or in the case of such solvents as water, ammonia, or hydrogen fluoride, through hydrogen bonding with the anion. In many cases the mechanism of the solvation process and the structure of the resulting solvate is unknown.

Solvation reactions in which water is the solvent are called hydration reactions, and if ammonia is the solvent the reactions is called ammoniates, respectively. Other solvates also take their names from the solvent involved, e.g., alcoholates, hydrazinates (from hydrazine) and etherates (from ether).

4

Chemistry in Protonic Solvents—Liquid Ammonia, Liquid Hydrogen Fluoride, Sulphuric Acid, Liquid Hydrogen Cyanide, Acetic Acid and Liquid Hydrogen Sulphide

INTRODUCTION

The solvent characteristics of water have already been discussed (See chapter 1) and it is merely included in this chapter for comparative purposes. The other compounds are all 'good' solvents and the first three are the most frequently used of all protonic non-aqueous solvents. It is clear from the list of physical properties of the protonic solvents (Table 4.1) that there will be a wide variation in their solution properties. The dielectric constant of liquid ammonia is considerably lower than that of the other three, all of which have dielectric constants of the magnitude that one might expect for very good solvents. In consequence salts of either doubly charged anions or cations tend to be insoluble in ammonia. Magnesium, calcium, strontium, and barium salts are insoluble as are carbonates and sulphates. Both sulphuric acid and liquid hydrogen fluoride are such strong acids that the number of true salts in these systems is very limited and discussion of solubilities is limited to the hydrogen sulphates and fluorides respectively.

A property of the solvents which it is worthwhile to compare is the equivalent ionic conductivity at infinite dilution of various ions (Table 4.2) is clear from the data that, whilst in water the H_3O^+ and OH^- ions are abnormally conducting as are the F^- ion in HF and the HSO_4^- and $H_3SO_4^+$ ions in sulphuric acid, there is no abnormal conduction shown by the NH_4^+ and NH_2^- ions in liquid ammonia.

Table 4.1 The Physical Properties of Water, Liquid Ammonia, Liquid Hydrogen Fluoride, and Sulphuric Acid

	NH_3	H_2O	HF	H_2SO_4	CH_3COOH
Melting point (°C)	−77.70	0	−89.37	10.371	16.7
Boiling point (°C)	−33.38	100	19.51	290.317	118
Liquid range (°C)	44.3	100	109	~300	102
Viscosity (centipoise)	0.2543 (−33.5 °C)	0.8904 (25 °C)	0.256 (0 °C)	24.54 (25 °C)	–
Density (g cm^{-3})	0.6900 (−40 °C)	1.00 (4 °C)	1.002 (0 °C)	1.8269 (25 °C)	–
Dielectric constant	23 (−33.4°)	78.5	84 (0 °C)	100 (25 °C)	7.1
Specific conductivity (ohm^{-1} cm^{-1})	$\sim 10^{-11}$	5×10^{-7}	$\sim 10^{-6}$ (0 °C)	1.0439×10^{-2}	–
Trouton's constant	23.28	26.0	24.7		–
Auto-protolysis constant	5.1×10^{-27}	10^{-14} (25 °C)	$\sim 2 \times 10^{-12}$	2.7×10^{-4} (25 °C)	–

Table 4.2 Equivalent Ionic Conductivities

	NH_3 (−33.5°C)	H_2O (25°C)	H_2SO_4 (25°C)	HF (0°C)
Na^+	158	50.1	~3	
NH_4^+	142	73.5		
H_3O^+		349.8		
NO_3^-	177	71.4		
NH_2^-	166	—		
OH^-		198.3		
Br^-	170	78.2		
K^+	177	73.5	~5	120
Cl^-		76.3		
F^-		55.4		280
HSO_4^-			151.2	
$H_3SO_4^+$			152	

LIQUID AMMONIA

Of the numerous non-aqueous solvents in use, the earliest to be systematically investigated as a solvent for chemical reaction was liquid ammonia. Liquid ammonia is now used routinely as the solvent for many types of organic and inorganic syntheses, and as a synthetic reagent in many such processes.

Physical Characteristics of Liquid Ammonia

The principal physical constant for ammonia are listed in Table 4.1.

A comparison of such physical constants for ammonia as melting point, boiling point, heat of vaporization and heat of fusion with the corresponding values for phosphine, arsine, and stibine reveals that the properties of ammonia are abnormal with respect to phosphine, arsine, and stibine in the same sense as the properties of water are out-of-line with respect to hydrogen sulfide, hydrogen selenide, and hydrogen telluride. The properties of hydrogen fluoride are similarly abnormal with respect to the other hydrogen halides.

The abnormal characteristics of water, ammonia, and hydrogen fluoride are readily explainable on the basis of intermolecular association by means of "hydrogen-bonding". Hydrogen bonds involve the formation of a bridge between two highly electronegative atoms by means of a proton. The nature of this bond is the subject of considerable discussion. Both an electrostatic model and a covalent model for the hydrogen bond have been presented but both of these fail to account for certain experimental observations. Hydrogen bonds are known to have bond energies of the order of a few kilocalories, which is of the right order of magnitude to have a great influence on the properties of solvents.

Ammonia molecules have a high degree of polarity resulting from the pyramidal structure of the molecule and the polarity of the N—H bond. The additional contribution resulting from the configuration of the charge distribution of the unshared electron pair must not, however, be neglected. It is believed that the bonding orbitals used by the nitrogen atom in uniting with three hydrogen atoms are sp^3 hybrids. The fourth orbital would, therefore, be an sp^3 hybrid also and would be directed in space in such a way that the negative charge density of the electron pair occupying that orbital would reinforce the dipole resulting from the three N—H bonds. The bond angle H—N—H (107°) is somewhat less than tetrahedral because of the "compressive" effect of the repulsion of the bonding electrons by the unshared pair. This repulsion is greater than that between the bonding pairs.

In the ammonia crystal each nitrogen atom has six neighboring nitrogen atoms at a distance of 3.380 ± 0.004 Å, this distance corresponding to that of a weak N—H . . . N bond. The energy of this bond, as estimated from the heat of sublimation of ammonia, and assuming a value of 180 kJ mol^{-1} for van der Waals forces, is approximately 54 kJ/mol.

Liquid ammonia undergoes autoionization in a manner analogous to that of water but, as its very low specific conductance indicates, the extent of autoionization is even less than that of water.

$$2NH_3 \rightleftharpoons NH_4^+ + NH_2^-$$

The ion-product constant, $K_{ion} = [NH_4^+][NH_2^-]$ is very low (1.9×10^{-33} at -50°C).

Solubilities in Liquid Ammonia

There have probably been more quantitative measurements on liquid ammonia than on any other non-aqueous solvent. It may, in general, be remarked that, as might be expected from the fact that is dielectric constant is much less than that of water, liquid ammonia is a much better solvent for organic compounds than is water. On the other hand, liquid ammonia is, in general, a poorer solvent for ionic substances than is water. There are, however, many ionic compounds which have a high degree of solubility in liquid ammonia.

In general salts containing doubly charged ions are insoluble in the solvent, probably because their lattice energies are too high. Salts with highly polarizable anions tend to be more soluble than those without.

Elements

Among the metallic elements, the alkali metals and alkaline earth metals are highly soluble in liquid ammonia, giving characteristic blue colored solutions. Europium and ytterbium, and possibly other rare earth metals, likewise form these solutions. Under special conditions liquid ammonia solutions of aluminium and beryllium are obtainable. These metal-ammonia solutions will be discussed in detail later.

The non-metals phosphorus, sulphur, and iodine, dissolve readily in liquid ammonia, but these solubility processes involve reaction with the solvent. For example, the solution process for sulphur has been shown to be

$$5/4\,S_8 + 16NH_3 \rightleftharpoons N_4S_4 + 6(NH_4)_2S$$

As indicated, this reaction is reversible and the evaporation of a freshly prepared liquid ammonia solution of sulphur yields free sulphur again.

Inorganic Compounds

There are few salts of multivalent anions which have appreciable solubility in liquid ammonia. Thus sulphates, sulphites, carbonates, phosphates, arsenates, oxides, and sulfides, are virtually insoluble in liquid ammonia. Metal hydroxides are likewise insoluble. All the metal amides except those of the alkali metals are insoluble. Lithium amide is also insoluble, and sodium amide is soluble only to the extent of 0.004 g. per 100 g. of NH_3. Potassium, rubidium, and cesium amides have a considerable solubility in liquid ammonia.

The metal salts which are, in general, most soluble in liquid ammonia include the thiocyanates, perchlorates, nitrates, nitrites, and many of the iodides. It is particularly interesting to note that in liquid ammonia the general order of solubility for halides is as follows:

$$\text{iodides} > \text{bromides} > \text{chlorides} > \text{fluorides}.$$

Most ammonium salts (except those of multivalent anions as mentioned above) are soluble in liquid ammonia, and some of these, such as the nitrate, acetate, and thiocyanate, are so highly soluble as to cause them to deliquesce in the presence of ammonia gas. Saturated solutions of these ammono-deliquescent salts have such a low vapor pressure of ammonia as to be stable at 0° C. A saturated solution of ammonium nitrate in liquid ammonia is stable at room temperature.

The solubilities of several inorganic salts in liquid ammonia at 25° C are listed in Table 4.3.

Table 4.3 Solubilities in Liquid Ammonia at 25°C (g./100 g.NH_3)

NH_4Cl	102.5	KNH_2	3.6
NH_4Br	237.9	KCl	0.04
NH_4I	368.4	KBr	13.5
NH_4SCN	312.0	KI	182.0
NH_4ClO_4	137.9	KCNO	1.70
NH_4NO_3	390.0	$KClO_3$	2.52
$(NH_4)_2S$	120.0	$KBrO_3$	0.002
$(NH_4)_2SO_3$	0.0	KI_3	0.0
$(NH_4)_2HPO_4$	0.0	KNO_3	10.4
$(NH_4)HCO_3$	0.0	K_2SO_4	0.0
$(NH_4)_2CO_3$	0.0	K_2CO_3	0.0
(NH_4) (CH_3COO)	253.2		
		AgCl	0.83
$LiNO_3$	243.66	AgBr	5.92
Li_2SO_4	0.0	AgI	206.84
		$AgNO_3$	86.04
$NaNH_2$	0.004		
NaF	0.35	$Ca(NO_3)_2$	80.22
NaCl	3.02	$Sr(NO_3)_2$	87.08
NaBr	137.95	$Ba(NO_3)_2$	97.22
NaI	161.9	$BaCl_2$	0.0
NaSCN	205.5		
$NaNO_3$	97.6	MnI_2	0.02
$Na_2S_2O_3$	0.17	ZnI_2	0.1
Na_2SO_4	0.0	ZnO	0.0
		H_3BO_3	1.92

Organic Compounds

In general, the solubilities of covalent organic compounds in liquid ammonia are considerably higher than in water, probably owing to the greater dispersion energy interaction. These data are summarized in Table 4.4.

Table 4.4 Solubilities of Organic Compounds in Liquid Ammonia

Hydrocarbons: Alkanes are insoluble. Alkenes and Alkynes have a slight solubility. Benzene is highly soluble. Toluene forms two liquid layers with liquid ammonia below 15°C.

Alcohols: Polyhydric alcohols and simple alcohols are miscible with liquid ammonia in all proportions. Hydroxy benzenes (phenol, etc.) are quite soluble.

Carboxylic Acids: Converted to ammonium salts. Those of low molecular weight acids are soluble. Solubility is less for the higher molecular weight compounds.

Esters: Simple esters are highly soluble in liquid ammonia, but as the alkyl groups increase in size, solubility decreases.

Aldehydes and Ketones: Moderately soluble but aldehydes and some ketones react with ammonia.

Ethers: Diethyl ether is moderately soluble but higher molecular-weight ethers are less so.

Alkyl Sulphuric and Alkyl and Aryl Sulphonic Acids: Converted to ammonium salts which are soluble in liquid ammonia.

Amines: Simple amines are quite soluble in liquid ammonia but solubility decreases with increasing molecular weight. Primary amines are more soluble than secondary amines which are more soluble than tertiary amines.

Acid Amides and Amidines: Simple members of these series are quite soluble.

Heterocyclic Nitrogen Bases: Pyridine, quinoline, indole, pyrrole, carbazole, simple triazoles, and simple tetrazoles are quite soluble in liquid ammonia.

SOLUTIONS OF METALS IN LIQUID AMMONIA

The most unusual feature of liquid ammonia as a solvent is its power to dissolve the alkali metals and, to a lesser extent, the alkaline earth metals and aluminium. The metals dissolve to give blue solutions which when dilute have identical absorption spectra whatever the metal. The blue colour is due to the short-wavelength tail of a broad band with a peak at approximately 15,000 Å. The solutions are extremely good conductors of electricity. Highly concentrated solutions (>1 M) are bronze coloured and have conductivities nearly as high as pure metals.

Instead of large positive temperature coefficients of conductivity, characteristic of electrolyte conduction, and owing to decrease of solvent viscosity with temperature, metal-ammonia solutions have temperature coefficients of nearly zero. Normal metals have a negative temperature coefficient of conductivity. All these metal-ammonia solutions are metastable and on long standing or in the presence of a suitable catalyst, such as iodine or ferric iodide, decompose to the amide and hydrogen:

$$2M + 2NH_3 \rightarrow H_2 + 2MNH_2$$

However, in clean apparatus with pure reagents the solutions can last for many weeks if kept at low temperatures.

The metal solutions are paramagnetic and the molar magnetic susceptibility of the solution approaches that of a mole of free electron spins, $N\mu_0^2/kT$, at infinite dilution. As the concentration is increased, however, the molar susceptibility decreases rapidly.

The conductivity of these solutions can be explained by assuming that in very dilute solutions the metal dissociates to give ammoniated cations and electrons:

$$M = M^+ + e^-$$

The electrons occupy cavities surrounded by ammonia molecules. Calculations based on the partial molar volumes of the solutes in these solutions and the partial molar volumes of K^+ and Na^+ ions give these cavities a radius of 3.34 Å. As the concentration of the solution is increased the M^+ and e^- species associate. Eventually incipient metallic behaviour occurs and the equivalent conductivity increases rapidly.

However, it is necessary to postulate other equilibria to explain the magnetic susceptibility data. These are:

$$2M^+ + 2e^- = M_2$$

and

$$M + e^- = M^-$$

An interesting aspect of several metal-ammonia systems is the fact that in certain temperature and concentration ranges the systems exist as two immiscible liquid phases in equilibrium. The heavier of these phases is blue and is the less concentrated in the metal; the lighter phase is the more concentrated in the metal and is bronze in color. This behavior has been reported for the lithium, sodium, potassium, calcium, strontium, and barium systems.

The solubilities of some of the alkali metals in liquid ammonia at various temperatures are given in Table 4.5.

Table 4.5 Solubilities of Li, Na, K, and Cs in Liquid NH_3

Lithium							
Temp., °C	0	−33.2	−39.4	−63.5	...	...	...
Molality	16.31	15.66	16.25	15.41	...	...	...
Sodium							
Temp., °C	22	0	−30	−33.5		−70	−100
Molality	9.56	10.00	10.63	10.93		11.29	11.79
Potassium							
Temp., °C	0	−33.2	−50	−100	−33.5	...	...
Molality	12.4	11.86	12.3	12.2	12.05	...	...
Cesium							
Temp., °C	−50	...	...	...	...	...	...
Molality	25.1	...	...	...	...	...	...

When solutions of alkali metals in liquid ammonia are evaporated the free metal is recovered. However, when ammonia solutions of calcium, strontium or barium are evaporated solid phases of composition $M(NH_3)_6$ are obtained. These are excellent electrical conductors and are metallic in appearance.

Blue solutions are also obtained when solutions of tetra-alkyl-ammonium halides in liquid ammonia are cathodically reduced.

Reactions of Metal-Ammonia Solutions

These solutions are strong reducing agents (stronger than hydrogen) and since many compounds are soluble in liquid ammonia the oxidation-reduction reaction is homogeneous. Reducing agents stronger than hydrogen liberate hydrogen from water and cannot generally be used in aqueous solution. Moreover, the high conductivity and strong colour of the metal-ammonia solutions mean that oxidation-reduction reactions using them can be followed both colorimetrically and conductimetrically.

The metal solutions are quickly decolorized by ammonium salts, the ammonium ion being reduced to ammonia and free hydrogen:

$$NH_4^+ + e^- \rightarrow NH_3 + \tfrac{1}{2}H_2$$

The metal solutions will also react with weakish acids, e.g. sulphamide, in liquid ammonia:

$$SO_2\begin{matrix}\diagup NH_2\\ \diagdown NH_2\end{matrix} + 2e^- \rightarrow SO_2\begin{matrix}\diagup NH^-\\ \diagdown NH^-\end{matrix} + H_2$$

The simple hydrides of germanium, arsenic and phosphorus also react to give the mono-sodium derivatives:

$$GeH_4 + e^- \rightarrow GeH_3^- + \tfrac{1}{2}H_2$$

$$PH_3 + e^- \rightarrow PH_2^- + \tfrac{1}{2}H_2$$

$$AsH_3 + e^- \rightarrow AsH_2^- + \tfrac{1}{2}H_2$$

Most reduction reactions involving metal-ammonia solutions can be considered to belong to one of the following three types:

1. *Electron Addition without Bond Cleavage*

$$X + e^- \rightarrow X^-$$

An example of this type of reaction is:

$$O_2 + e^- + O_2^-$$

and this offers a method for the preparation of pure samples of alkali metal superoxides. Further reaction between the superoxide ion and excess of metal-ammonia solutions can take place:

$$O_2^- + e^- \rightarrow O_2^{2-}$$

With nitrite ion:

$$NO_2^- + e^- \rightarrow NO_2^{2-}$$

Complex transition metal ions also can be reduced by metal-ammonia solutions:

$$MnO_4^- + e^- \rightarrow MnO_4^{2-}$$
$$Ni(CN)_4^{2-} + 2e^- \rightarrow Ni(CN)_4^{4-}$$
$$Pt(NH_3)_4^{2+} + 2e^- \rightarrow Pt(NH_3)_4$$

2. *Bond Cleavage by the Addition of One Electron*

The reaction with the ammonium ion falls into this category:

$$NH_4^+ + e^- \rightarrow NH_3 + \tfrac{1}{2}H_2$$

as do the reactions with other hydrides:

$$AsH_3 + e^- \rightarrow AsH_2^- + \tfrac{1}{2}H_2$$

and with ethanol:

$$C_2H_5.OH + e^- \rightarrow C_2H_5.O^- + \tfrac{1}{2}H_2$$

and with very weak acids.

The reaction with organic sulphides is of the same type:

$$R_2S + e^- \rightarrow RS^- + R\cdot \rightarrow RS^- + \tfrac{1}{2}R_2$$

Occasionally a stable radical is formed:

$$(C_2H_5)_3SnBr + e^- \rightarrow (C_2H_5)_3Sn\cdot + Br^-$$

3. *Bond Cleavage with the Addition of Two or More Electrons*

Examples of this are:

$$Ge_2H_6 + 2e^- \rightarrow 2GeH_3^-$$
$$N_2O + 2e^- \rightarrow N_2 + O^{2-} \xrightarrow{NH_3} N_2 + OH^- + NH_2^-$$
$$NCO^- + 2e^- \rightarrow NC^- + O^{2-} \xrightarrow{NH_3} NC^- + OH^- + NH_2^-$$

The reactions of the elements with metal-ammonia solutions probably fall into this class:

$$S_8 + ne^- \rightarrow S_2^{2-} \text{ and other polysulphides}$$
$$Se + ne^- \rightarrow Se_2^{2-} \text{ and polyselenides}$$

Reactions in Liquid Ammonia

Chemical reactions which occur in liquid ammonia may be classified as 1. acid-base reactions, 2. amphoteric reactions, 3. metathetic reactions, 4. ammonolytic reactions, 5. ammoniation reactions and 6. redox reactions.

1. *Acid-Base reactions*

The autoionisation of liquid ammonia is just like that of water.

$$2NH_3 \rightleftharpoons NH_4^+ + NH_2^-$$
$$2H_2O \rightleftharpoons H_3O^+ + OH^-$$

Ammonium ion, NH_4^+, in liquid ammonia is the counterpart of hydronium ion, H_3O^+, in aqueous solution and amide ion, NH_2^-, that of hydroxyl ion, OH^-. Hence, ammonium salts which yield ammonium ions in liquid ammonia behave as acids and metal amides which yields amide ions act as bases and, when mixed, the two neutralise one another with the formation of ammonia, just as in aqueous solution acids and bases neutralise one another with the formation of water.

Examples of neutralisation reactions in liquid ammonia are:

$$NH_4Cl + KNH_2 \rightarrow KCl + 2NH_3$$
$$2NH_4NO_3 + PbNH_2 \rightarrow Pb(NO_3)_2 + 3NH_3$$

The acidic nature of ammonium salts in liquid ammonia is further evident from the liberation of hydrogen from reactions of these solutions with some active metals such as Na, K and Ca.

In liquid ammonia

$$Na + 2NH_4Cl \rightarrow H_2 + 2NaCl + 2NH_2$$

In water

$$2Na + 2HCl \rightarrow H_2 + 2NaCl$$

Amphoteric behaviour is shown in both solvents. Many metallic amides and imides are amphoteric like many metallic hydroxides and oxides. For example, zinc amide dissolves in liquid ammonia solution of potashamide, forming $K_2[Zn(NH_2)_4]$ and zinc hydroxide in a like manner dissolves in aqueous solution of potassium hydroxide yielding potassium zincate, $K_2[Zn(OH)_4]$.

In liquid ammonia

$$ZnCl_2 + 2NH_3 \rightarrow Zn(NH_2)_2 + 2KCl^-$$
$$Zn(NH_2)_2 + 2KNH_2^- \rightarrow [Zn(NH_2)_4]^{2-}$$

In water

$$ZnCl_2 + 2KOH^- \rightarrow Zn(OH)_2 + 2KCl^-$$
$$Zn(OH)_2 + 2KOH \rightarrow K_2[Zn(OH)_4]^{2-}$$

All acids which are strong acids in water react completely with ammonia (i.e., are "leveled") to form ammonium ions:

$$HClO_4 + NH_3 \rightarrow NH_4^+ + ClO_4^-$$
$$HNO_3 + NH_3 \rightarrow NH_4^+ + NO_3^-$$

Acids which behave as weak acids in water (with pK_a up to about 12) react completely with ammonia and hence are strong acids in this solvent.

$$HC_2H_3O_2 + NH_3 \rightarrow NH_4^+ + C_2H_3O_2^-$$

Molecules showing no acidic behavior at all in water may behave as weak acids in ammonia:

$$NH_2CONH_2 + NH_3 \rightleftharpoons NH_4^+ + NH_2CONH^-$$

Thus ammonia levels all species showing significant acidic behaviour (in aqueous solution) and enhances the acidity of very weakly acidic species.

Most species that would be considered bases in water are either insoluble or behave as weak based in ammonia. Extremely strong bases, however, may be 'leveled' to the amide ion and behave as strong bases:

$$H^- + NH_3 \rightarrow NH_2^- + H_2\uparrow$$
$$O^{2-} + NH_3 \rightarrow NH_2^- + OH^-$$

2. *Metathetical* (*Precipitation*) *Reaction*

These reactions in liquid ammonia proceed in the same manner as in aqueous medium. When two ions capable of forming a sparingly soluble self react, they form a precipitate in both the media. However, owing to the differences in solubility between the two solvents, the direction of the reaction in one case may be opposite of the other. For example:

In water

$$BaCl_2 + 2AgNO_3 \rightarrow Ba(NO_3)_2 + 2AgCl \downarrow$$
$$KCl + AgNO_3 \rightarrow KNO_3 + AgCl \downarrow$$

In ammonia

$$2AgCl + Ba(NO_3) \rightarrow 2AgNO_3 + BaCl_2 \downarrow$$
$$AgCl + KNO_3 \rightarrow AgNO_3 + KCl \downarrow$$

However, solubility of salts in liquid ammonia frequently differs from that in water. Consequently, precipitation reactions which appear curious at first sight, often take place in liquid ammonia. Thus, silver bromide and barium nitrate react in liquid ammonia to yield a precipitate of barium nitrate but a reverse reaction occurs in aqueous medium.

Some other examples of the precipitation reactions that can be carried out in liquid ammonia medium but not possible in aqueous medium include the following:

$$2NH_4I + Zn(NO_3)_2 \rightarrow ZnI_2 \downarrow + 2NH_4NO_3$$
$$2NH_4Br + Sr(NO_3)_2 \rightarrow SrBr_2 \downarrow + 2NH_4NO_3$$
$$Ba(NO_3)_2 + 2AgBr \rightarrow BaBr_2 \downarrow + 2AgNO_3$$

3. *Ammonolytic Reactions*

The solvolysis reactions in liquid ammonia are known as ammonolytic reactions and are similar to hydrolysis reaction.

$$AlCl_3 + NH_3 \rightarrow Al(NH_2)_2^{2+} + H^+ + 3Cl^-$$
$$AlCl_3 + H_2O \rightarrow Al(OH)^{2+} + H^+ + 3Cl^-$$

Certain covalent metallic and non-metallic halides may be completely ammonolysed by liquid ammonia just as they are completely hydrolysed by water.

$$TiCl_4 + 8NH_3 \rightarrow Ti(NH_2)_4 + 4NH_4Cl$$
$$TiCl_4 + 4H_2O \rightarrow Ti(OH)_4 + 4HCl$$

The halides of non-metals react with water yielding hydrolytic products in which some or all the halogens are replaced by hydroxyl groups. Similar reactions take place with ammonia.

$$SiCl_4 + 8NH_3 \rightarrow Si(NH_2)_4 + 4NH_4^+ + 4Cl^-$$

$$SiCl_4 + 4H_2O \rightarrow Si(OH)_4 + 4H^+ + 4Cl^-$$
$$PCl_5 + 10NH_3 \rightarrow P(NH_2)_5 + 5NH_4^+ + 5Cl^-$$
$$PCl_5 + 4H_2O \rightarrow H_3PO_4 + 5H^+ + 5Cl^-$$
$$BCl_3 + 3NH_3 \rightarrow B(NH_2)_3 + 3NH_4^+ + 3Cl^-$$
$$BCl_3 + 3H_2O \rightarrow B(OH)_3 + 3H^+ + 3Cl^-$$

4. Ammonation Reactions

Ammonia forms a large number of metal ammines by the ammonation of metal salts in liquid ammonia or by the action of ammonia on the anhydrous metal salts. Typical examples include $Cr(NH_3)_6Cl_3$, $Co(NH_3)_6Cl_2$ and $Ni(NH_3)_6Cl_2$. These ammines are formal analogus of the corresponding salt hydrates, $Cr(H_2O)_6Cl_3$, $Co(H_2O)_6Cl_3$ and $Ni(H_2O)_6Cl_2$. The two groups of compounds, however, differ in many respects.

5. Redox Reactions

Strong oxidising agents do not exist in liquid ammonia. Reducing agents, on the other hand, exhibit enhanced activity. Dilute solutions of alkali metals in liquid ammonia have a deep blue colour. They are excellent electrical conductors strongly paramagnetic and are noted for their powerful reducing properties. To explain the above properties it has been suggested that alkali metal dissolve in liquid ammonia forming solvated cations and free electrons.

$$M + xNH_3 \rightarrow M(NH_3)_x^+ + e^- \text{ (M = alkali metal)}$$

Many compounds which could not be prepared earlier by reduction in aqueous solution have now been prepared by carrying out reduction in metal ammonia solution. An interesting example of the use of these solutions is for the preparation of the complex cyanides, $K_4[M(CN)_4]$ (where M = Ni, Co or Pd) in which the metal is in the zero oxidation state. Another example is the preparation of carbonyl hydrides.

Some example of redox reactions are:

$$4Na + 2NH_3 \rightarrow 2NaOH + 2NaNH_2$$
$$2Na + 2NH_4Br \rightarrow 2NaBr + 2NH_3 + H_2$$
$$K + KMnO_4 \rightarrow K_2MnO_4$$
$$2K + N_2O + NH_3 \rightarrow 2K^+ + NH_2^- + OH^- + N_2$$

LIQUID HYDROGEN FLUORIDE

Investigations of this solvent in the past have been handicapped by its reactivity with glass and quartz. However, the advent of fluorine-containing plastics such as Teflon (polytetrafluoroethylene) and Kel-F (polychlorotrifluoroethylene) and the use of copper and stainless-steel vacuum lines have made quantitative studies easier. Most salts dissolve in the solvent with reaction and most non-ionic compounds with protonation. Even when simple fluorides are dissolved, bifluorides rather than fluorides are recovered on removal of the solvent and the fluoride ion must be heavily solvated in solution. Unlike ammonia the solvent is not particularly effective in solvating cations.

The solubilities of a range of fluorides in liquid hydrogen fluoride are given in Table 4.6. It will be seen from the table that salts with small cations are less soluble than those with large ones and that solubility decreases dramatically with increasing cation charge.

Table 4.6 The Solubilities of Fluorides in Liquid Hydrogen Fluoride

Fluoride	*Solubility*	*Fluoride*	*Solubility*
LiF	10.3	BeF_2	0.015
NaF	30.4	MgF_2	0.025
KF	36.5 (8°C)	CaF_2	0.87
RbF	110.0 (20°C)	SrF_2	14.83
CsF	199.0 (10°C)	BaF_2	5.60
NH_4F	32.6 (17°C)	CuF_2	0.010
AgF	83.2 (19°C)	AgF_2	0.048
TlF	580.0	PbF_2	2.62
AlF_3	<0.002	NiF_2	0.037
CeF_3	0.043	FeF_2	0.006
TlF_3	0.081	CrF_2	0.036
SbF_3	0.536	HgF_2	0.54
BiF_3	0.010	NbF_5	6.8
CeF_4	0.10	TaF_5	15.2
ThF_4	<0.006	SbF_5	Miscible in all proportions

Data are at 12°C and Solubilities are in g/100 g of HF

It is likely that the solubility of NbF_5, TaF_5, and SbF_5 is due to their functioning as fluoride ion acceptors in the solvent. Chlorine trifluoride and bromine trifluoride are completely miscible with hydrogen fluoride and are probably acting as fluoride ion donors:

$$ClF_3 + HF \rightarrow ClF_2^+ + HF_2^-$$

The solubilities of organic covalent compounds in liquid hydrogen fluoride are extremely high and in many cases the solutions have a very high conductivity, indicating that the solute has been protonated. The reason for high solubilities is thus very different from that for ammonia. Solubilities are summarized in Table 4.7.

Solvates are formed exclusively by hydrogen bonding and the only ones reported are fluorides: KHF_2, KH_2F_3, KH_4F_5, and $H_3O^+H_3F_4^-$.

Table 4.7 Solubilities of Organic Compounds in Liquid Hydrogen Fluoride

Hydrocarbons Saturated aliphatic hydrocarbons and their halogenated (other than fluoro) derivatives are insoluble. Aromatic hydrocarbons, such as benzene and the methylbenzenes tend to be soluble. Aromatic and aliphatic compounds are soluble if they carry substituents containing nitrogen, oxygen, or sulphur atoms.

In general the presence of electron-withdrawing groups, such as halogen or nitro on an aromatic ring, has the effect of lowering solubility. Thus, phenols are soluble, mono-nitrophenols possess only limited solubility, and trinitrophenol is insoluble. Note, however, that nitrobenzene is more soluble than benzene.

Butadiene and other unsaturated compounds polymerize.

Alcohols Aliphatic alcohols are miscible in all proportions.

Amines They are protonated to produce extremely soluble salts, as are heterocylic nitrogen bases.

Esters and ethers Protonated and very soluble,

Carboxylic acids Protonated and very soluble. Acetic acid is miscible in all proportions.

Chemical Reactions in Liquid Hydrogen Fluoride

Liquid HF has a liquid range from 184 to 292 K.

Although hydrogen fluoride has a rather low specific conductance, its high dielectric constant makes it an excellent ionising solvent. It is regarded as one of the most water-like of all non-aqueous solvents. It dissolves many inorganic and organic compounds to yield highly conducting solutions. Inorganic compounds, in general, are more soluble in this solvent than the organic compounds.

The disadvantages of hydrogen fluoride as a solvent and reaction medium are its ability to dissolve only relatively few substances without chemical reaction and its poisonous character.

The dissolution of a solute in hydrogen fluoride is believed to proceed via one of the following mechanisms:

(*a*) *Ionizing Dissociation*

It may dissociate into ions just as in other ionising solvents. This behaviour is exemplified by the solvent action on potassium fluoride.

$$KF + HF \rightleftharpoons K^+ + HF_2^-$$

(*b*) *Chemical Reaction*

$$KCN + 2HF \rightleftharpoons HCN + K^+ + HF_2^-$$

or

Chemical reactions involving more than simple replacement may take place, such as

$$H_2SO_4 + 2HF \rightleftharpoons HOSO_2F + H_3O^+ + F^-$$

Simple dissolution of a solute in hydrogen fluoride is less frequent than other reactions. Consequently, very few salts dissolve in liquid hydrogen fluoride to yield their simple ions.

1. *Acid-Base Reactions*

The autoionisation of hydrogen fluoride may be represented as

$$3HF \rightleftharpoons H_2F^+ + HF_2^- \text{ or } F^-.HF$$

Thus any substance capable of increasing the concentration of H_2F^+ ion will act as an acid whereas any substance increasing the concentration of the fluoride ion, F^- will act as a base. There are only a few compounds which are capable of donating protons readily to hydrogen fluoride. Thus, a number of substances which behave as strong acids in aqueous solution behave as based in this solvent, for example, nitric acid and sulphuric acid.

$$HNO_2 + HF \rightarrow H_2NO_3^+ + F^-$$

$$H_2SO_4 + HF \rightarrow H_3SO_4^+ + F^-$$

Perchloric acid, a strong acid in aqueous medium, is amphoteric in this solvent.

$$HClO_4 + HF \rightarrow H_2ClO_4^+ + F^-$$ (HClO, acts as a base)

$$HClO_4 + HF \rightarrow H_2F^+ + ClO_4^-$$ ($HClO_4$ acts as a weak acid).

The only compounds which seem to act as acids are certain Lewis acids notably BF_3, AsF_3, PF_5 and SbF_5.

$$BF_3 + 2HF \rightarrow H_2F^+ + BF_4^-$$

$$SbF_5 + 2HF \rightarrow H_2F^+ + SbF_6^-$$

Substances acting as bases in hydrogen fluoride include ionic fluorides. The fluorides increase the concentration of fluoride ion and thus shifting the auto-ionisation equilibrium in hydrogen fluoride to the left and decreasing the concentration of H_2F^+.

$$3HF \rightleftharpoons H_2^+F + HF_2^-$$

$$KF \rightleftharpoons K^+ + F^-$$

Weak acids in other systems behave as based in hydrogen fluoride. For example acetic acid behaves as a base in hydrogen fluoride.

$$CH_3COOH + HF \rightarrow CH_3COOH_2^+ + F^-$$

A neutralization reaction in HF may be represented by the following example

$$KF + SbF_5 \rightleftharpoons KSbF_6$$

Amphoteric behaviour

Aluminium fluoride is relatively insoluble in liquid hydrogen fluoride, but dissolves readily on the addition of sodium fluoride:

$$AlF_3 + NaF \rightarrow Na^+ + AlF_4^-$$

The reaction is slightly more complicated than this as the actual nature of the aluminium fluoroanions present depends on the concentration of sodium fluoride. On addition of boron trifluoride, aluminium fluoride is reprecipitated:

$$NaAlF_4 + BF_3 \rightarrow AlF_3\downarrow + NaBF_4$$

Potassium hexafluorochromate(III) is soluble in hydrogen fluoride, with precipitation of CrF_3. The precipitate redissolves in excess of sodium fluoride:

$$CrF_3 + 3NaF \rightarrow NaCrF_6$$

The addition of boron trifluoride causes reprecipitation of the simple fluoride:

$$Na_3CrF_6 + 3BF_3 \rightarrow CrF_3 \downarrow + 3NaBF_4$$

Many metallic chlorides, bromides and iodides react with hydrogen fluoride forming the fluorides and liberating the appropriate hydrogen halide.

$$KCl + HF \rightleftharpoons HCl + KF$$

2. Redox Reactions

Oxidation-reduction reactions between hydrofluo acids and metals in hydrogen fluoride are not much different from those between nitric acid and metals in the aqueous system. The reactions illustrating the similarity are:

$$3HAsF_6 + 2Ag \rightarrow 2AgAsF_6 + As^{III}F_3 + 3HF$$
$$7HIF_6 + 6Ag \rightarrow 6AgIF_6 + HI + 6HF$$
$$4HNO_3 + 3Ag \rightarrow 3AgNO_3 + NO + 2H_2O$$

Metathetical Reaction: Some of the precipitation reactions which may occur in liquid hydrogen fluoride are

$$NaSO_4 + 2AgF \rightarrow Ag_2SO_4 \downarrow + 2NaF$$
$$NaClO_4 + AgF \rightarrow AgClO_4 \downarrow + NaF$$
$$KIO_4 + AgF \rightarrow AgIO_4 \downarrow + KF$$

Of the many preparations which have been carried out in liquid hydrogen fluoride, perhaps the most important one is that of silver tetrafluoroborate. It is obtained as a precipitate when silver nitrate and boron trifluoride solutions in hydrogen fluoride are mixed.

$$AgNO_3 + BF_3 + 2HF \rightarrow AgBF_4 + H_2NO_3F$$

3. Solvolysis Reactions

Most simple salts are solvolysed, e.g.:

$$KCN + HF \rightarrow HCN \uparrow + K^+ + F^-$$
$$KCl + HF \rightarrow HCl \uparrow + K^+ + F^-$$

The initial solvolysis may be followed by a second reaction, e.g.:

$$KNO_3 + HF \rightarrow HNO_3 + K^+ + F^-$$
$$HNO_3 + HF \rightarrow H_2NO_3^+ + F^-$$
$$H_2NO_3^+ + HF \rightarrow NO_2^+ + H_3O^+ + F^-$$

$$K_2SO_4 + 2HF \rightarrow H_2SO_4 + 2K^+ + 2F^-$$
$$H_2SO_4 + HF \rightarrow HSO_3F + H_2O$$
$$H_2O + HF \rightarrow H_2O^+ + F^-$$

Acid halides and anhydrides are completely solvolysed, e.g.:

$$CH_3.C\begin{matrix}\nearrow O \\ \searrow Cl\end{matrix} + HF \rightarrow HCl\uparrow + CH_3.\ CH_3.C\begin{matrix}\nearrow O \\ \searrow F\end{matrix}$$

$$(CH_3.CO)_2O + 2HF \rightarrow CH_3.C(OH)_2^+ + CH_3.CH_3.C\begin{matrix}\nearrow O \\ \searrow F\end{matrix}$$

Metathetical reactions. These are of minor significance in the solvent because of the extensive solvolysis which most other anions undergo in the solvent. When either HCl, HBr, or HI gas is passed through a solution of AgF or TlF in liquid hydrogen fluoride, then silver or thallium chloride, bromide, or iodide is precipitated. However, on standing without passing the gas the silver or thallium salt slowly redissolves, indicating that the first reaction is dependent on a pressure of HCl, HBr, or HI.

Silver borofluoride is precipitated when BF_3 is passed through a solution of silver fluoride in liquid hydrogen fluoride.

Electrochemical Oxidations in Liquid Hydrogen Fluoride

Because of the very high potential needed for the anodic reaction in liquid hydrogen fluoride:

$$F^- \rightleftharpoons \tfrac{1}{2}F_2 + e^-$$

the solvent is particularly suited to the performance of anodic oxidation reactions. These usually proceed by the insertion of fluorine in a molecule and have been widely used commercially as well as in the research laboratory for the production of fluorine-containing organic compounds. Among the inorganic compounds, the production of NFH_2, NF_2H, and NF_3 by the electrolysis of ammonium fluoride in liquid hydrogen fluoride represents the only convenient synthetic route to these difficultly available compounds. Trifluoroacetic acid is readily obtained by the electrolysis of acetic acid in liquid HF.

When the organic starting compounds is insoluble (e.g. an aliphatic hydrocarbon), soluble organic or inorganic additives are introduced to increase the conductivity and speed of electrolysis. Table 4.7 lists the products of anodic oxidation of a variety of compounds in liquid HF.

Table 4.7 Production of Anodic Oxidation in Liquid HF

Reactant	*Products*
NH_4F	NF_3, NHF_2, NH_2F
H_2O	OF_2
SCl_2, SF_4	SF_6
$NaClO_4$	ClO_4F
$(CH_3)_2S$ or CS_2	CH_3SF_5, $(CF_3)_2SF_4$
$(C_2H_5)_2O$	$(C_2F_5)_2O$
$CH_3.NH_2$	$CF_3.NF_2$
$(CH_3)_2NH$	$(CF_3)_2NF$
$(CH_3)_3N$	$(CF_3)_3N$
$CH_3.CN$	$CF_3.CN$, $C_2F_5.NF_2$

Solutions of Compounds of Biological Interest

Anhydrous hydrogen fluoride is a powerful solvent for polysaccharides and proteins. Cellulose dissolves freely to form conducting solutions. The material recovered from such solutions, designated a glucosan, yields glucose on mild hydrolysis.

Not only are the water-soluble proteins readily soluble in the solvent, but many fibrous proteins such as silk fibroin and collagen also dissolve freely. Although chemical reactions may occur in some cases, in many cases biological activity is retained. Insulin can be recovered from the solvent with full retention of biological activity and ribonuclease and lysozyme retain their enzymatic action after dissolution at low temperatures and recovery. The iron containing proteins cytochrome *c* and haemoglobin dissolve in liquid hydrogen fluoride to form solutions with absorption spectra similar to those in water, as do the metal phthalocyanines. Vitamin B_{12} forms a deep olive-green colour in hydrogen fluoride as contrasted to its normal deep red; however, it survives solution and can be recovered with full biological activity.

SULPHURIC ACID

Sulphuric acid is perhaps the most extensively studied of all the strongly acidic solvents. It has a high dielectric constant and can be conveniently handled in glass apparatus at room temperature. Its only disadvantage is as a preparative solvent. If a compound is not precipitated from the solvent then removal of sulphuric acid is difficult because of its high boiling point.

The pioneer work in sulphuric acid as a solvent was carried out by A. Hantzsch during the first decade of this century.

Major advances in this study resulted from the work of L. Hammett in the early 1930's. During the past, however, our knowledge of chemistry in this solvent has been greatly expanded.

Sulphuric acid is a very good solvent for numerous organic compounds, many of which are able to accept protons, i.e., act as bases, in it. Because of its strongly acidic character, sulphuric acid may be used as a solvent in which very weak bases, such as ketones and aromatic nitro-compounds, are basic enough to act as proton acceptors toward sulphuric acid molecules.

Physical Properties of Sulphuric Acid

Some of the more important physical constants for sulphuric acid are listed in Table 4.8.

Table 4.8 Physical Constants of Sulphuric Acid

Freezing Point	10.371 °C
Boiling Point	290–317 °C
Density, $d_4^{25°}$	1.8269
Viscosity (25 °C)	24.54 centipoise
Heat Capacity	1.413 J/deg./g.
Heat of Fusion	107 kJ/mole
Dielectric Constant (10 °C)	120.
Dielectric Constant (25 °C)	100.
Specific Conductance (25 °C)	0.010439 $ohm^{-1}cm^{-1}$

Pure sulphuric acid is prepared by adding aqueous sulphuric acid to fuming sulphuric acid (anhydrous sulphuric acid containing excess sulphur trioxide) until a maximum freezing point of 10.371°C is obtained. Sulphuric acid is, as its high viscosity and high boiling point indicate, a highly associated solvent. Study of the crystalline solid indicates a layer structure in which each sulphuric acid molecule is hydrogen bonded to four other such molecules, the length of these bonds being 2.85 Å. It may reasonably be assumed that this association persists to a large extent in the liquid state.

The sulphuric acid molecule has the tetrahedral structure indicated in Fig. 4.1. This structure may be rationalized on the basis of the use by the sulphur atom of sp^3 hybrid orbitals in the formation of bonds with the four oxygen atoms. Since *d* orbitals are also available on the sulphur atom and unshared electrons on the oxygen atoms, it

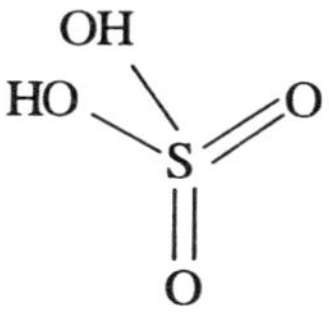

OH
HO — S = O
‖
O

Fig. 4.1 The structure of the H_2SO_4 molecule.

is probable that the sulphur to oxygen bond have considerable π bond character.

Acids in the solvent give the sulphuric acidium ion, $H_3SO_4^+$, and bases the hydrogen sulphate ion, HSO_4^-. Because of its high acidity, few acids are known in the solvent and bases are much the largest class of electrolyte. Like liquid hydrogen fluoride it has a levelling effect on the strengths of bases. Despite its high acidity, sulphuric acid is also appreciably basic and this is shown by its very large auto-protolysis constant, K_{ap}, for $[H_3SO_4^+][HSO_4^-]$ which has a value of 2.7×10^{-4}.

In addition to auto-protolysis:

$$2H_2SO_4 = H_3SO_4^+ + HSO_4^-$$

there are other dissociation equilibria which are a consequence of a primary dissociation into water and sulphur trioxide:

$$H_2SO_4 = H_2O + SO_3$$

Since water is a base in the solvent:

$$H_2O + H_2SO_4 = H_3O^+ + HSO_4^-$$

this reaction proceeds extensively to the right. Sulphur trioxide forms disulphuric acid:

$$SO_3 + H_2SO_4 = H_2S_2O_7$$

which is partly ionized as an acid:

$$H_2S_2O_7 + H_2SO_4 = H_3SO_4^+ + HS_2O_7^-$$

Since the ions $H_3SO_4^+$ and HSO_4^- are in equilibrium because of the auto-protolysis reaction, the ions H_3O^+ and $HS_2O_7^-$ must also be in equilibrium:

$$2H_2SO_4 = H_3O^+ + HS_2O_7^-$$

This has been called the ionic self-dehydration reaction, with an equilibrium constant, K_{id}. The value of the equilibrium constants are given in Table 4.9. The total molar concentration of species other than H_2SO_4 (i.e. HSO_4^-, $H_3SO_4^+$, H_3O^+, $HS_2O_7^-$, $H_2S_2O_7$, and H_2O) is 0.0424 M at 25 °C.

Table 4.9 Self-dissociation Reactions in Sulphuric Acid

Equation	*Equilibrium constant expression*	*Value at* 25 °C
$2H_2SO_4 = H_3SO_4^+ + HSO_4^-$	$K = [H_3SO_4^+][HSO_4^-]$	2.7×10^{-4}
$2H_2SO_4 = H_3O^+ + HS_2O_7^-$	$K = [H_3O^+][HS_2O_7^-]$	5.1×10^{-5}
$H_2S_2O_7 + H_2SO_4 = H_3SO_4^+ + HS_2O_7^-$	$K = \frac{[H_3SO_4^+][HS_2O_7^-]}{[H_2S_2O_7]}$	1.4×10^{-2}
$H_2O + H_2SO_4 = H_3O^+ + HSO_4^-$	$K = \frac{[H_3O^+][HSO_4^-]}{[H_2O]}$	1

At atmospheric pressure, 100% sulphuric acid has a freezing point of 10.371 °C. The 100% acid is most easily prepared by adding dilute oleum to slightly aqueous sulphuric acid until the maximum freezing point of 10.3718 °C is reached.

The conductivities of solutions in sulphuric acid also offer a very powerful means of determining the nature of the reaction between the solute and the solvent. The mobilities of the $H_3SO_4^+$ and HSO_4^- ions are so very much higher than those of any other ions in the solvent that the conductivities of solutions of acids and bases in sulphuric acid are determined almost entirely by the concentrations of $H_3SO_4^+$ and HSO_4^- respectively. Consequently, γ, the number of moles of HSO_4^- ions or the number of moles of $H_3SO_4^+$ ions produced by one mole of an electrolyte, can be obtained from the conductivities of any acid or base.

Some correction needs to be made at low concentrations for the repression of the solvent auto-protolysis. This is most conveniently done by comparing the concentration of a standard monohydrogen sulphate, such at $KHSO_4$, needed to produce a given conductivity with the concentration of the electrolyte under examination required to give the *same* conductivity. The ratio of the two is γ.

Table 4.10 Solubilities of Metal Sulphates in Sulphuric Acid at 25°C

Sulphate	*Solubility (mole %)*	*Sulphate*	*Solubility (mole %)*
Li_2SO_4	14.28	$PbSO_4$	0.12
Na_2SO_4	5.28	$CuSO_4$	0.08
K_2SO_4	9.24	$FeSO_4$	0.17
Ag_2SO_4	9.11	$NiSO_4$	very small
$MgSO_4$	0.18	$HgSO_4$	0.78
$CaSO_4$	5.16	Hg_2SO_4	0.02
$BaSO_4$	8.85	$Al_2(SO_4)_3$	<0.01
$ZnSO_4$	0.17	$Tl_2(SO_4)_3$	<0.01

Solubilities in Sulphuric Acid

Because of the strong hydrogen bonding between sulphuric acid molecules, it is difficult for a solute to break down the solvent structure and dissolve unless the solute is highly solvated, that is, ionic. Because of its high dielectric constant sulphuric acid is an excellent solvent for electrolytes; however, its high acidity means that most solutes dissolve with reaction. Table 4.10 gives the solubilities of a range of metal sulphates in the solvent. Those that are appreciably soluble are, of course, recovered as the bisulphate.

Most soluble sulphates are recovered from sulphuric acid as bisulphates, but this can hardly be classed as solvate formation. However, solid phases containing more sulphuric acid than this can often be prepared at low temperatures. Thus at 25°C the solid phases in equilibrium with saturated solutions in sulphuric acid are: $2LiHSO_4, H_2SO_4$; $4NaHSO_4, 7H_2SO_4$; $KHSO_4, H_2SO_4$; $Mg(HSO_4)_2$,-$2H_2SO_4$; $Ca(HSO_4)_2, 2H_2SO_4$; $Ba(HSO_4)_2$, $2H_2SO_4$, and there is presumably strong hydrogen bonding between the bisulphate ion and the sulphuric acid in these crystal lattices.

Conductances in Sulphuric Acid

Mobilities of various cations and anions in 100% sulphuric acid would, because of the high viscosity of their solvent, be expected to be quite low and this is, in general, true. Table 4.11

Table 4.11 Ion Mobilities at Infinite Dilution in Sulphuric Acid and in Water at 25°C

Ion	H_2SO_4	H_2O
Na^+	~3	50.1
K^+	~5	73.5
Ba^{++}	~2	63.6
H_3O^+	~5	349.8
OH^-	. . .	198.6
$H_3SO_4^+$	242	. . .
HSO_4^-	171	. . .

However, the $H_3SO_4^+$ ion and the HSO_4^- ion have very high mobilities in sulphuric acid. In fact, it is because of the high mobilities of these two kinds of ions, as well as because of the considerable degree of ionic dissociation of sulphuric acid, that this solvent has quite an appreciable electrical conductance in spite of its high viscosity.

Because of these very high mobilities of $H_3SO_4^+$ and HSO_4^- the electrical conductances of various acids and bases in 100% sulphuric acid are principally determined by the number of $H_3SO_4^+$ or HSO_4^- ions per mole of the particular acid or base in the sulphuric acid solution. Thus, measurement of electrical conductances provides valuable information concerning the ionization processes for certain electrolytes.

Chemical Reactions in Sulphuric Acid

Reactions in sulphuric acid are protonation, solvolysis, oxidation, or dehydration reactions. Sometimes a mixture of reactions occurs.

Acid-Base reactions: There are very few strong acids in the sulphuric acid solvent system; even $H_2S_2O_7$ is relatively weak. Both conductivity and cryoscopic measurements are employed to determine the nature of the reactions.

Many aqueous acids behave as bases in sulphuric acid, e.g.:

$$CH_3.COOH + H_2SO_4 = CH_3.C(OH)_2^+ + HSO_4^-$$
$$H_3PO_4 + H_2SO_4 = P(OH)_4^+ + HSO_4^-$$

The behaviour of nitric acid is slightly more complex:

$$HNO_3 + 2H_2SO_4 = NO_2^+ + H_3O^+ + 2HSO_4^-$$

Presumably the protonated form of the acid, $H_2NO_3^+$, breaks down to give NO_2^+ and H_2O, which is subsequently protonated. The behaviour of hydrochloric acid in the solvent is more complex. If a chloride is dissolved in sulphuric acid there is an evolution of HCl gas, but at low concentrations in the cold HCl reacts quantitatively to give chlorosulphuric acid:

$$HCl + 2H_2SO_4 = HClSO_3 + H_3O^+ + HSO_4^-$$

In some cases the protonated form of a carboxylic acid, $R.C(OH)_2^+$, like $H_2NO_3^+$, can lose water and give $R.CO^+$. Thus:

$$R.COOH + 2H_2SO_4 = R.CO^+ + H_3O^+ + 2HSO_4^-$$

An example of an acid which undergoes this type of ionization to yield a stable acyl ion is mesitoic acid, $(CH_3)_3C_6H_2.COOH$, which gives $(CH_3)_3C_6H_2.CO^+$. On dilution with water, mesitoic acid is recovered and on dilution with methanol the methylester is obtained.

A number of carboxylic acids are unstable in sulphuric acid and decompose to give carbon monoxide, presumably via the acyl ion, $R.CO^+$:

$$R.CO^+ = R^+ + CO$$

Examples are formic acid:

$$H.COOH + H_2SO_4 = CO + H_3O^+ + HSO_4^-$$

and oxalic acid:

$$(COOH)_2 + H_2SO_4 = CO + CO_2 + H_3O^+ + HSO_4^-$$

Strong acids are produced in sulphuric acid by dissolving either boric acid or boric oxide in the solvent:

$$H_3BO_3 + 6H_2SO_4 = B(HSO_4)_4^- + 3H_3O^+ + 2HSO_4^-$$
$$B_2O_3 + 9H_2SO_4 = 2B(HSO_4)_4^- + 3H_3O^+ + HSO_4^-$$

Solutions of the free acid can be prepared by dissolving boric acid or boric oxide in oleum instead of in sulphuric acid, in which case the H_3O^+ ion is removed by the reaction:

$$H_3O^+ + SO_3 = H_3SO_4^+$$

Acid-Base Titrations. A number of acid-base titrations have been carried out in 100% sulphuric acid solution. Such titrations are most conveniently followed conductometrically. This is possible because the $H_3SO_4^+$ and HSO_4^- ions have much higher conductances in sulphuric acid than any other cations or anions. Hence, a minimum conductance occurs at or very near the equivalence point in the neutralization reaction:

$$H_3SO_4^+ + HSO_4^- \rightarrow 2H_2SO_4$$

If either the acid or base involved in the titration is only partially dissociated (i.e., is a weak acid or base) the minimum point will not occur at the equivalence point, and corrections will have to be made. In general, it has been demonstrated that for an acid-base titration in 100% sulphuric acid, the molar ratio of base to acid (n_b/n_a) at the minimum conductance at 25°C is given by the equation

$$n_b/n_a = 0.98\ (1 + 0.017/K_b)\ (1 + 0.014/K_a)$$

where K_b and K_a refer to the ionization constants of the respective base and acid.

A solution of $HB(HSO_4)_4$ can be titrated conductimetrically with a solution of a strong base such as $KHSO_4$.

Both tin tetra-acetate and lead tetra-acetate dissolve in sulphuric acid to give acids:

$$Sn(OAc)_4 + 8H_2SO_4 = Sn(HSO_4)_6^{2-} + 4CH_3.C(OH)_2^+ + 2HSO_4^-$$
$$Pb(OAc)_4 + 8H_2SO_4 = Pb(HSO_4)_6^{2-} + 4CH_3.C(OH)_2^+ + 2HSO_4^-$$

Neither of these acids is very strong and in the case of hexa(hydrogen sulphato) plumbic acid the variation of ν and γ with concentration in the above reaction has been used to determine the two dissociation constants of the acid:

$$H_2Pb(HSO_4)_6 + H_2SO_4 = H_2SO_4^+ + HPb(HSO_4)_6^-$$

$$K_1 = 1.2 \times 10^{-2} \text{ mol kg}^{-1}$$

$$HPB(HSO_4)_6^- + H_2SO_4 = H_3SO_4^+ + Pb(HSO_4)_6^{2-}$$

$$K_2 = 1.8 \times 10^{-3} \text{ mol kg}^{-1}$$

and the acid is of about the same strength as $H_2S_2O_7$. A summary of the strengths of various acids in sulphuric acid is given in Table 4.12.

Table 4.12 Acid Strengths in Sulphuric Acid

Acid	*Ionization constant (mol kg^{-1})*
$H_2S_2O_7$	1.4×10^{-2}
HSO_3F	3×10^{-3}
$H_2Pb(HSO_4)_6$	1.2×10^{-2}
$HPb(HSO_4)_6^-$	1.8×10^{-3}
$HB(HSO_4)_4$	4×10^{-1}
$HClO_4$	very weak
HSO_3Cl	very weak

Protonation

Most ketones, aldehydes, carboxylic acids, ethers, amines, and amides are strong bases in sulphuric acid, as are organic phosphines. However, in some cases further reaction can occur. Amides appear to protonate on the oxygen rather than on the nitrogen:

$$R.CO.NH_2 + H_2SO_4 = R.C(OH).NH_2^+ + HSO_4^-$$

and this protonation is followed by a slow solvolysis:

$$R.C(OH).NH_2^+ + 2H_2SO_4 = R.CO_2H_2^+ + NH_4^+ + HS_2O_7^-$$

Ethers are monoprotonated in sulphuric acid:

$$R.O.R' + H_2SO_4 = RR'OH^+ + HSO_4^-$$

but again the reaction is followed by a slow solvolysis:

$$RR'OH^+ + 2H_2SO_4 = RHSO_4 + R'HSO_4 + H_3O^+$$

When triphenylmethanol is dissolved in sulphuric acid it ionizes according to the equation:

$$(C_6H_5)_3C.OH + 2H_2SO_4 = (C_6H_5)_3C^+ + H_3O^+ + 2HSO_4^-$$

giving a stable yellow solution.

Stable carbonium ions can also be prepared by dissolving some substituted alkenes in sulphuric acid:

$$(C_6H_5)_2C{:}CH_2 + H_2SO_4 = (C_6H_5)_2C.CH_3^+ + HSO_4^-$$

A number of inorganic and organic compounds function as weak bases in sulphuric acid. Thus, selenium dioxide is soluble in sulphuric acid to give a bright yellow solution, and in dilute solutions it behaves as a weak base:

$$SeO_2 + H_2SO_4 = HSeO_2^+ + HSO_4^-$$

though in view of the solubility the equilibrium probably ought to be written:

$$SeO_2 + H_2SO_4 = SeO(OH).HSO_4 = SeO.OH^+ + HSO_4^-$$

Some organic compounds, notably nitro compounds, sulphones, and sulphoxides, are weak bases in sulphuric acid. Surprisingly, so are nitriles, the base strengths of which one would have expected to be greater, though here protonation is complicated by slow solvolysis:

$$R.CN + H_2SO_4 = R.CNH^+ + HSO_4^-$$

$$R.CNH^+ + HSO_4^- = R.CO.NH_2 + SO_3$$

The ionization constants of some weak bases are given in Table 4.13.

Table 4.13 Ionization Constants of Weak Bases in Sulphuric Acid

Base	$K_{\text{ionization}}$
$CH_3.NO_2$	2.5×10^{-3}
$C_6H_5.NO_2$	1.0×10^{-2}
$p.MeC_6H_4.NO_2$	9.6×10^{-2}
$CH_3.CN$	1.6×10^{-1}
$C_6H_5.CN$	7.0×10^{-2}
$(C_6H_5)_2SO$	1.6×10^{-2}
SeO_2	4.4×10^{-3}

Liquid Hydrogen Cyanide

Liquid hydrogen cyanide is a highly associated through hydrogen bonding. It has the highest dielectric constant and dipole moment of any other water-like solvent. On the basis of the above physical properties. We would expect liquid hydrogen cyanide to be the best of all ionising solvents. But it is not so. In comparison to water, it is undoubtedly an inferior solvent and dissolves mainly the covalent compounds.

1. *Acid-base Reactions*

Hydrogen cyanide autoionizes as:

$$2HCN \rightleftharpoons H_2CN^+ + CN^-$$

Thus the protonic acids such as HCl, H_2SO_4, HNO_3 etc. act as acids and can be neutralized against alkali cyanide solution which act as bases. Such acid-base reactions can be followed conductometrically.

$$\underset{\text{acid}}{HCl} + \underset{\text{base}}{KCN} \rightarrow \underset{\text{salt}}{KCl} + \underset{\text{solvent}}{HCN}$$

2. *Solvolysis Reactions*

Some salts such as Ag_2SO_4 undergo solvolysis reactions.

$$Ag_2SO_4 + 2HCN \rightleftharpoons AgCN + H_2SO_4$$

$$CuCl_2 + 2HCN \rightleftharpoons Cu(CN)_2 + 2HCl$$

3. *Metathetical Reactions*

Only a few alkali metal salts such as NaCl, KCl, NH_4Cl, $NaNO_3$, K_2SO_4 and KI dissolve and ionize in hydrogen cyanide. Hence not many metathetical reactions are possible; some examples are

$$2NaCl + K_2SO_4 \rightarrow Na_2SO_4\downarrow + 2KCl$$

$$AgI + NaCl \rightarrow NaI + AgCl$$

Acetic Acid

Acetic acid is an associated solvent and exists as a dimer in the liquid state. It has a conveniently wide liquid range 289.6 to 391 K. It is non-ionic and stable under ordinary conditions. A most serious disadvantage of this solvent is the difficulty involved in the preparation of anhydrous acid. It has a dipole moment of zero and a low value of dielectric constant (9.7). From these properties acetic acid would be expected to be a poor solvent for ionic compounds but surprisingly large number of such compounds are soluble in this solvent.

1. *Acid-base Reactions*

Autoionization of acetic acid takes place as follows

$$2CH_3COOH \rightleftharpoons CH_3COOH_2^+ + CH_3COO^-$$

Thus substances giving the solvated proton ($CH_3COOH_2^+$) in solution may be regarded as acids and those yielding the acetate ion may be considered as bases. Typical acid-base reactions occur between soluble acetates and strong acids.

$$\underset{\text{base}}{CH_3COONa} + \underset{\text{acid}}{HCl} \rightarrow \underset{\text{salt}}{NaCl} + \underset{\text{solvent}}{CH_3COOH}$$

It is not possible to titrate weak organic bases like acetamide and acetanilide in aqueous medium but they may be titrated in acetic acid medium. Perchloric acid, the most highly dissociated of the strong acids in acetic acid, is generally employed as the titrant. The neutralisation reaction is followed potentiometrically and also with indicators. The weak organic bases function as quite strong bases by virtue of the reaction

$$\underset{\substack{\text{Weak}\\ \text{organic base.}}}{B} + CH_3COOH \rightleftharpoons BH^+ + CH_3COO^-$$

2. *Metathetical Reactions*

Of the many precipitation reactions which take place in this solvent, some are:

$$2AgNO_3 + CaCl_2 \rightarrow 2AgCl + Ca(NO_3)_3$$
$$Pb(C_2H_3O_2)_2 + 2KCl \rightarrow PbCl_2 + 2KC_2H_3O_2$$
$$Cu(NO_3)_2 + 2NaC_2H_3O_2 \rightarrow Cu(C_2H_3O_2)_2 + 2NaNO_3$$
$$BaI_2 + 2NaNO_3 \rightarrow Ba(NO_3)_2\downarrow + 2NaI$$

Certain compounds, such as zinc acetate, exhibit amphoteric behaviour in this medium.

$$Zn^{2+} + 2C_2H_3O_2^- \rightarrow \underset{\text{insoluble}}{Zn(C_2H_3O_2)_2}$$

$$Zn(C_2H_3O_2)_2 + 2C_2H_3O_2^- \rightarrow \underset{\text{Soluble}}{[Zn(C_2H_3O_2)_4]^{2-}}$$

3. *Complex Formation Reactions*

Certain complex formation reactions occur in acetic acid just like in aqueous medium. For example, Fe^{3+} and thiocynate ions react in this medium to form the red coloured complex, $Fe(CNS)^{2+}$. The same complex is produced in aqueous solution from interaction of the above ions.

4. *Solvolytic Reactions*

Solvolytic reactions occuring in acetic acid medium are exemplified by the following:

$$SO_2Cl_2 + 4CH_3COOH \rightarrow SO_2(CH_3COO)_2 + 2CH_3C(OH)_2 + 2Cl^-$$

$$CN^- + CH_3COOH \rightarrow CH_3COO^- + HCN$$

Liquid Hydrogen Sulphide

Liquid hydrogen sulphide behaves more like an organic solvent than like water and is thus a much poorer solvent for ionic compounds than water. Only a few inorganic compounds dissolve in this solvent without chemical reaction.

Liquid hydrogen sulphide possesses a very narrow and inconveniently liquid range (188–212 K) for experimental work and is highly poisonous.

The low conductance of the pure compound precludes the existence of any appreciable autoinisation and the development of the corresponding system of acids and bases. The self-ionization scheme postulated is

$$2H_2S \rightleftharpoons H_3S^+ + SH^-$$

Some acid-base reactions have been observed in liquid hydrogen sulphide. For example:

$$\underset{\text{base}}{[N(C_2H_5)_3H]^+SH^-} + \underset{\text{acid}}{HCl} \rightarrow \underset{\text{solvent}}{[N(C_2H_5)_3H)^+Cl^-} + \underset{\text{salt}}{H_2S}$$

A number of compounds get solvolysed in liquid hydrogen sulphide. Some example of solvolytic reactions which occur in this medium are given below:

$$PCl_5 + H_2S \rightleftharpoons PSCl_3 + 2HCl$$

$$SbCl_5 + H_2S \rightleftharpoons SbSCl_3 + 2HCl$$

$$HgCl_2 + H_2S \rightleftharpoons HgS + 2HCl$$

$$2AsCl_3 + 3H_2S \rightleftharpoons As_2S_3 + 6HCl$$

$$SnCl_4 + H_2S \rightleftharpoons SnS_2 + 4HCl$$

5

Non-Protonic Solvents—Liquid Dinitrogen Tetroxide, Liquid Sulphur Dioxide and Liquid Halides

INTRODUCTION

A molecule of a protonic solvent is capable, under one set of circumstances, of accepting a proton, and under other conditions of donating a proton to another molecule or ion. We shall now consider solvents, which contain no hydrogen atoms. It is desirable to use the Lewis definition of acids and bases (acids are electron acceptors; bases are electron donors) in discussing chemical reactions in these solvents since the Bronsted definitions cannot be applied to substances which do not contain hydrogen.

In this chapter liquid oxide solvents, dinitrogen tetroxide and sulphur dioxide and liquid halide solvents (both ionizing as well as covalent) are discussed.

LIQUID OXIDE SOLVENTS

Liquid Dinitrogen Tetroxide

The physical properties of N_2O_4 are listed in Table 5.1. Dinitrogen tetroxide melts at $-12.3°C$ (the triple point is at $-11.2°C$) and its normal boiling point is 21.3°C. Thus its liquid range is convenient for its use as a solvent. The liquid may be readily supercooled and has been cooled as low as $-110°C$ without it crystallizing. Its critical temperature is 158.2°C and its critical pressure is 100.0 atm. The density of dinitrogen tetroxide is 1.49 $g.cm^{-3}$ at 0°C. The electrical conductance of liquid dinitrogen tetroxide is very low; the specific conductance at various temperatures expressed in $ohm^{-1}\ cm^{-1}$ is given by the equation $\log_{10}\kappa = (-1267/T) - 8.260$. This corresponds to a value for κ of 2.36×10^{-13} ohm^{-1} cm^{-1} at 17°C.

Table 5.1 Physical Properties of Liquid N_2O_4

Melting point (°C)	–12.3
Boiling point (°C)	21.3
Liquid range (°C)	33.6
Density (g cm^{-3})	1.49 (0°C)
Dielectric constant	2.42 (18°C)
Specific conductivity (ohm^{-1}cm^{-1})	2.36×10^{-13} (17°C)

As would be expected from its very low dielectric constant, N_2O_4 is a poor solvent for ionic substances, solubilities being similar to solubilities in ether. Many organic solutes dissolve in N_2O_4 without change. Alkanes, aromatic hydrocarbons, halo and nitro compounds, and carboxylic acids are readily soluble. Many oxygen and nitrogen compounds are also soluble, for example ethers, esters and tertiary amines, and oxygen- and nitrogen-containing heterocyclic compounds. Many other organic compounds containing an active hydrogen atom, such as primary and secondary amines, alcohols, and ketones, dissolve with reaction.

Structure of Dinitrogen Tetroxide

X-ray diffraction studies of crystalline dinitrogen tetroxide indicate the following coplanar structure:

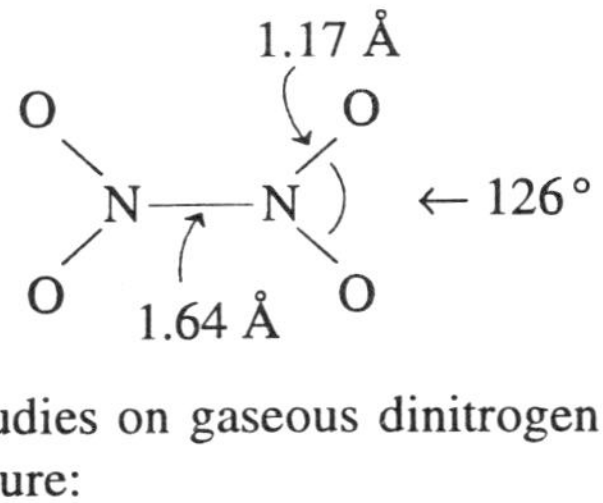

1.17 Å
O O
N—N ← 126°
O O
1.64 Å

Electron diffraction studies on gaseous dinitrogen tetroxide indicate a similar coplanar structure:

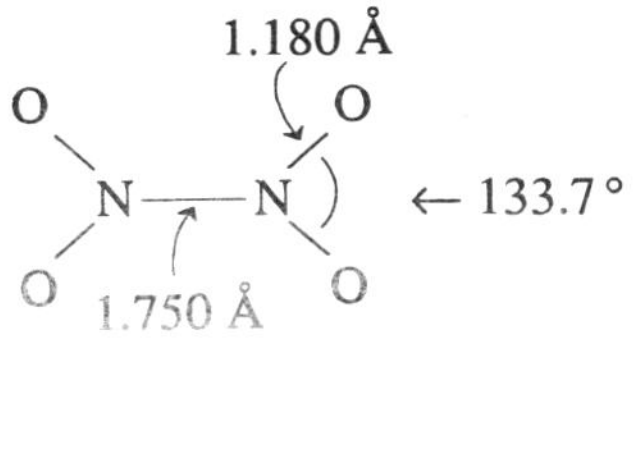

1.180 Å
O O
N—N ← 133.7°
O O
1.750 Å

It may be noted that in each case the nitrogen to nitrogen bond distance is considerably larger than is normal for a nitrogen to nitrogen single bond. This may be rationalized if we assume that the N_2O_4 molecule is a resonance hybrid of such valence bond structures as:

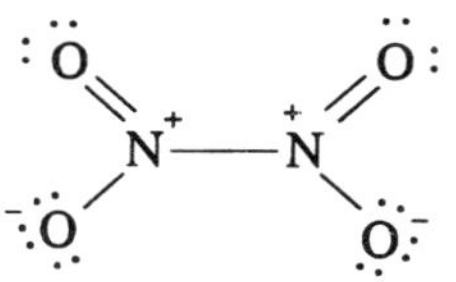

in which the adjacent nitrogen atoms bear positive charges. These would cause the nitrogen to nitrogen bond to be elongated.

It is of interest to note that recent studies of the infrared spectra of solid dinitrogen tetroxide indicate the existence in detectable amounts at low temperatures of two additional forms, (a) and (b), of dinitrogen tetroxide:

(a) 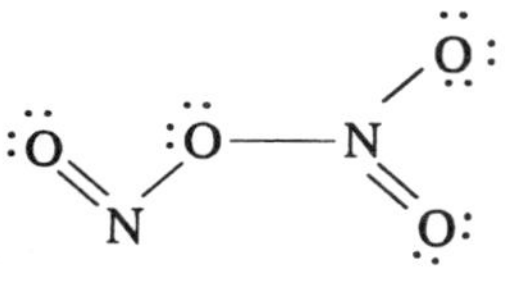 (hybridized with similar resonance forms)

(b) 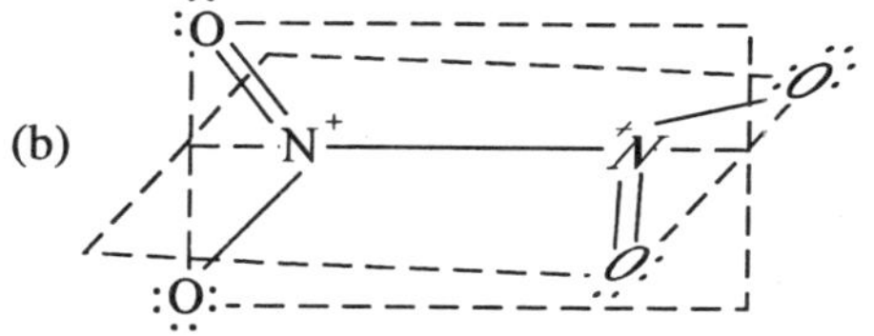

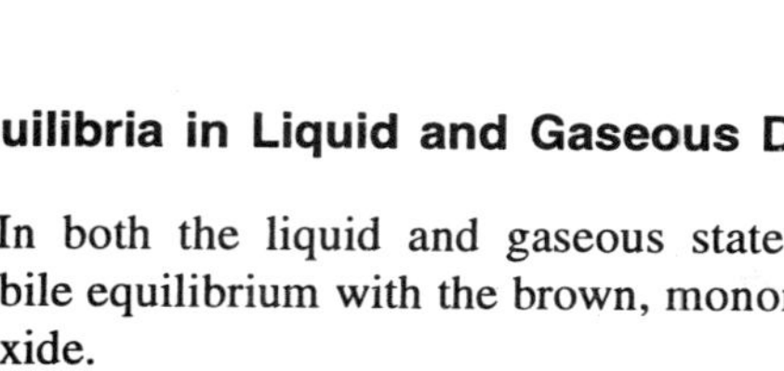

(hybridized with similar resonance forms) (the two NO_2 groups are in planes which are at right angles to each other).

Equilibria in Liquid and Gaseous Dinitrogen Tetroxide

In both the liquid and gaseous states dinitrogen tetroxide is in mobile equilibrium with the brown, monomeric species, NO_2, nitrogen dioxide.

$$N_2O_4 \rightleftharpoons 2NO_2$$

Since nitrogen dioxide contains an odd number of electrons and is, therefore, paramagnetic, the change in equilibrium with change in temperature may be readily followed by measuring the magnetic susceptibility of the equilibrium mixture. Because of the differences in the absorption spectra of the monomeric and dimeric species, the equilibrium can also be studied spectrophotometrically. Results obtained by these two methods indicate that the equilibrium constant $K = \frac{[NO_2]^2}{[N_2O_4]}$ for the reaction $N_2O_4 \rightleftharpoons 2NO_2$ in the liquid phase is of the order of 1.0×10^{-6} g cm^{-3} at 20°C. This means that the concentration of NO_2 in the liquid phase is quite low. However, in the vapour phase the concentration of NO_2 is 16.1% at the boiling point and increases rapidly as the temperature is raised.

Studies of the chemistry of liquid dinitrogen tetroxide demonstrate there is an additional equilibrium in the liquid phase, viz. the autoionization equilibrium

$$N_2O_4 \rightleftharpoons NO^+ + NO_3^-$$

The fact that this ionization occurs to only an exceedingly small extent is demonstrated by the very low magnitude of the specific electrical conductance of pure liquid dinitrogen tetroxide.

Certain chemical reactions of liquid dinitrogen tetroxide with organic compounds indicate also the probability that this substance is at least potentially capable of dissociating in accordance with the equilibrium

$$N_2O_4 \rightleftharpoons NO_2^+ + NO_2^-$$

The extent of this dissociation in pure liquid dinitrogen tetroxide is undoubtedly very small.

The proposed auto-ionization reaction in the liquid:

$$N_2O_4 \rightleftharpoons NO^+ + NO_3^-$$

must occur to an exceedingly small extent since the specific electrical conductivity of the purified liquid is so low.

Thus, compounds giving the NO^+ group would be acids in the solvent and substances giving NO_3^- ions would be bases; compounds such as NOCl and NOBr will behave as acids, and nitrates such as $Et_4N^+NO_3^-$ will behave as bases. A solution of NOCl in liquid N_2O_4 will react with solid $AgNO_3$:

$$NOCl + AgNO_3(s) \xrightarrow{N_2O_4(l)} AgCl(s) + N_2O_4$$

but the same reaction will proceed equally well if NOCl is dissolved in benzene or petroleum ether.

A wide variety of transition metal nitrates form solvates with N_2O_4. Some of these are listed in Table 5.2, together with their probable structures.

Table 5.2 Solvates with N_2O_4

Salt	*Solvate*	*Probable formulation*
$Zn(NO_3)_2$	$Zn(NO_3)_2, 2N_2O_4$	$(NO^+)_2[Zn(NO_3)_4^{2-}]$
$Fe(NO_3)_3$	$Fe(NO_3)_3, N_2O_4$	$NO^+[NO_3)_4^-]$
$Cu(NO_3)_2$	$Cu(NO_3)_2, N_2O_4$	$NO^+[Cu(NO_3)_3^-]$
$UO_2(NO_3)_2$	$UO_2(NO_3)_2, N_2O_4$	$NO^+[UO_2(NO_3)_3^-]$

Molecular Addition Compounds

A wide range of covalent compounds, particularly organic Lewis bases, also form adducts with N_2O_4. Boron trifluoride forms a complex N_2O_4, $2BF_3$ which has the structure:

```
          OBF3⁻
         /
NO2⁺ N
         \
          OBF3
```

an indication of an unusual ionization of the solvent molecule. A list of the adducts formed between N_2O_4 and organic molecules is given in Table 5.3. It will be seen from this table that the stability of these adducts varies greatly, and the fact that many of them are 1:1 adducts between the electron pair base and N_2O_4 is an indication that their structures may be described as:

```
O         O
  \      /
   N — N
  /  ↗   \
O  B      O
```

With stronger bases, including possibly the sulphoxides and certainly the tertiary amines, the structure is ionic. Thus, trimethyl-amine forms the salt $(CH_3)_3NNO^+NO_3^-$.

Salts such as alkylammonium halides have a sufficiently low lattice energy to be soluble. They undergo solvolysis in the solvent:

$$Et_4NCl + N_2O_4 \rightarrow NOCl + Et_4NNO_3$$

Table 5.3 Adducts of N_2O_4 with Organic Molecules

Organic compound	*Adduct composition*	*Melting point* (°C)	*Stability*
$(C_2H_5)_2O$	$2(C_2H_5)_2O, N_2O_4$	−74.8	Very low
Tetrahydrofuran	$(CH_2)_4O_2, N_2O_4$	−20.5	Moderate
$(CH_3)_2NNO$	$2(CH_3)_2NNO, N_2O_4$	+3	Moderate
$CH_3 \cdot COOH$	$2CH_3 \cdot COOH, N_2O_4$	+2	Moderate
$C_6H_5 \cdot COOC_2H_5$	$2C_6H_5 \cdot COOC_2H_5, N_2O_4$	−13	Moderate
$(CH_3)_2CO$	$2(CH_3)_2CO, N_2O_4$	−40	Low
$(CH_3)_2SO$	$(CH_3)_2SO, N_2O_4$	+38	Stable
$(C_2H_5)_2SO$	$(C_2H_5)_2SO, N_2O_4$	+14	Stable
C_6H_6	C_6H_6, N_2O_4	−7	Moderate
Mesitylene	$C_6H_3(CH_3)_3, N_2O_4$	−18	Low
$C_6H_5 \cdot CN$	$C_6H_5 \cdot CN, N_2O_4$	−26	Low

The degree of solvolysis depends on the solubility of the chloride in the solvent. Other solvolysis reactions are:

$$Mg(OH_2)_6^{2+}(Cl^-)_2 + 2N_2O_4 \rightarrow Mg(OH_2)_6^{2+}(NO_3^-)_2 + 2NOCl$$
$$Mg(ClO_4)_2 + 2N_2O_4 \rightarrow Mg(NO_3)_2 + 2NOClO_4$$

Acid-Base Reactions

The dissociation equilibrium mentioned above,

$$N_2O_4 \rightleftharpoons NO^+ + NO_3^-$$

was considered to be analogous to the autoionization reactions known or postulated for various other solvents, for example,

$$2H_2O \rightleftharpoons H_3O^+ + OH^-$$
$$2NH_3 \rightleftharpoons NH_4^+ + NH_2^-$$

According to this idea, substances furnishing the NO^+ group would be typical acids in dinitrogen tetroxide and substances furnishing NO_3^- ions would be typical bases. Such substances as NOCl and NOBr should behave as acids and the nitrates, e.g., $[Et_2NH_2]\,NO_3$ should act as bases in this system. The low dielectric constant of dinitrogen tetroxide does not favour ionic reactions. However, a solution of NOCl in liquid dinitrogen tetroxide reacts with solid silver nitrate

$$NOCl + AgNO_{3\,(s)} \underset{N_2O_4}{\overset{liq.}{\rightleftharpoons}} AgCl_{\,(s)} + N_2O_4$$

This reaction would be a typical neutralization reaction analogous to the following reactions from the water and ammonia systems:

$$(H_3O)Cl + NaOH \xrightarrow{H_2O} NaCl + 2H_2O$$

$$(NH_4)Cl + KNH_2 \xrightarrow[NH_3]{liq} KCl + 2NH_3$$

Solvolytic Reactions

A number of solvolytic reactions in liquid dinitrogen tetroxide have been investigated. Thus the salt diethyl ammonium chloride undergoes solvolysis in accordance with the following equation:

$$[Et_2NH_2]Cl + N_2O_4\ (l) \rightarrow NOCl + [Et_2NH_2]NO_3$$

The general type reaction $MCl + N_2O_4 \rightarrow NOCl + MNO_3$ is fully reversible in liquid dinitrogen tetroxide depending upon the solubility of MCl in the system, and the removal of NOCl from the system. The process for preparing NOCl from potassium chloride is in some ways analogous. In this process gaseous dinitrogen tetroxide is passed through a column of potassium chloride moistened with a trace of water. The following reaction occurs:

$$KCl_{(slightly\ moist)} + N_2O_4\ (g) \rightarrow NOCl\ (g) + KNO_3\ (s)$$

Other solvolytic reactions in liquid dinitrogen tetroxide include the following:

$$[Mg(H_2O)_6]Cl_2 + 2N_2O_4 \rightarrow [Mg(H_2O)_6](NO_3)_2 + 2NOCl$$
$$Mg(ClO_4)_2 + 2N_2O_4 \rightarrow Mg(NO_3)_2 + 2NO(ClO_4)$$

In the case of aluminium chloride, a mixture of $Al(NO_3)_3$ and $Al(NO_3)_3 \cdot N_2O_4$ is obtained. As indicated below, the latter compound is probably best formulated as $NO[Al(NO_3)_4]$.

In the presence of a small amount of water, lithium carbonate is solvolyzed in accordance with the equation:

$$Li_2CO_3 + 2N_2O_4 \rightarrow 2LiNO_3 + N_2O_3 + CO_2$$

Liquid dinitrogen tetroxide likewise reacts with calcium oxide, calcium carbonate, sodium carbonate, sodium hydroxide, zinc carbonate, and zinc sulphide to give anhydrous nitrates, sometimes admixed with nitrites. The reactions are, in many cases, quite slow and may sometimes be speeded up by adding small quantities of water. Commonly one can do this and still obtain an anhydrous nitrate product.

Reactions with Metals

Very active metals react with liquid dinitrogen tetroxide in a manner analogous to their reactions with water.

$$M(s) + N_2O_4(l) \rightarrow NO + MNO_3 \text{ (when M is an alkali metal)}$$

$$M(s) + H_2O(l) \rightarrow \tfrac{1}{2}H_2 + MOH$$

Just as the addition of hydrogen chloride to water increases its reactivity toward metals so also the addition of NOCl to liquid dinitrogen tetroxide increases its reactivity toward metals. Thus, metals such as zinc, iron, and tin react with solutions of NOCl in liquid dinitrogen tetroxide as follows.

$$M(s) + 2NOCl \xrightarrow[N_2O_4]{liq} MCl_2 + 2NO \quad (M = Zn, Fe, Sn, etc.)$$

analogous to

$$M(s) + 2HCl \xrightarrow{H_2O} MCl_2 + H_2$$

Amphoteric metals such as zinc behave toward liquid dinitrogen tetroxide solutions in a manner analogous to their behaviour toward aqueous acids and bases. Thus, zinc metal reacts only very slowly with liquid dinitrogen tetroxide. However, zinc reacts rapidly with solutions of NOCl or $[Et_2NH_2]NO_3$ in this solvent.

$$Zn + x[Et_2NH_2]NO_3 + 2N_2O_4(l) \rightarrow [Et_2NH_2]_x[Zn(NO_3)_{x+2}] + 2NO$$

Diethyl Ammonium Nitratozincate
(exact composition unknown)

analogous to

$$Zn + 2NaOH + 2H_2O \rightarrow Na_2[Zn(OH)_4] + H_2$$

$$Zn + 2KNH_2 + 2NH_3 \rightarrow K_2[Zn(NH_2)_4] + H_2$$

Zinc nitrate dissolves readily in solutions of diethyl ammonium nitrate in liquid dinitrogen tetroxide to yield a nitrato zinc complex.

$$Zn(NO_3)_2(s) + x[Et_2NH_2]NO_3 \xrightarrow[N_2O_4]{liq} [Et_2NH_2]_x[Zn(NO_3)_{x+2}]$$

This reaction is analogous to the following:

$$Zn(OH)_2(s) + 2NaOH \xrightarrow{H_2O} Na_2[Zn(OH)_4]$$

$$Zn(NH_2)_2(s) + 2KNH_2 \xrightarrow[NH_3]{liq.} K_2[An(NH_2)_4]$$

$$Al_2(SO_3)_3(s) + 3(Me_4N)_2SO_3 \xrightarrow[SO_2]{liq.} 2[Me_4N]_3[Al(SO_3)_3]$$

Copper metal reacts with a solution of dinitrogen tetroxide in ethyl acetate to yield a compound having the formula $Cu(NO_3)_2 \cdot N_2O_4$. Zinc nitrate reacts with liquid dinitrogen tetroxide to give the solvate $Zn(NO_3)_2 \cdot 2N_2O_4$. Uranyl nitrate, $UO_2(NO_3)_2$, likewise forms the solvate $UO_2(NO_3)_2 \cdot$N2O4 . The solvate $Fe(NO_3)_3 \cdot N_2O_4$ has also been prepared. These solvates may be formulated as the complex salts $NO[Cu(NO_3)_3]$, $(NO)_2[Zn(NO_3)_4]$, $NO[UO_2(NO_3)_3]$ and $NO[Fe(NO_3)_4]$.

LIQUID SULPHUR DIOXIDE

The physical properties of liquid sulphur dioxide are listed in Table 5.4. In general, covalent substances are considerably more soluble than ionic compounds in liquid sulphur dioxide. However, its higher dielectric constant means that it is a much better solvent for ionic compounds than liquid N_2O_4.

Table 5.4 Some Physical Properties of Sulphur Dioxide

Melting point (°C)	−75.46
Boiling point (°C)	−10.02
Liquid range (°C)	65.5
Viscosity of liquid (millipoise)	4.285 (−10°C)
Dielectric constant	15.4 (°C)
Specific conductivity ($ohm^{-1} cm^{-1}$)	$3–4 \times 10^{-8}$ (−10 °C)
Trouton's constant	22.7
Dipole moment(Debye)	1.62
Enthalpy of fusion ($kcal\ mol^{-1}$)	1.97
Enthalpy of vaporization ($kcal\ mol^{-1}$)	5.96

Structure of the Sulphur Dioxide Molecule

The structure of the sulphur dioxide molecule in the gaseous state has been shown by electron diffraction to correspond to the following diagram:

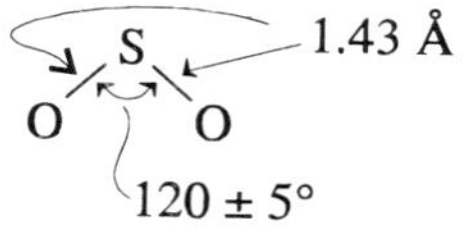

S, O, O — 1.43 Å — 120 ± 5°

This structure may be rationalized as a resonance hybrid of the following structures:

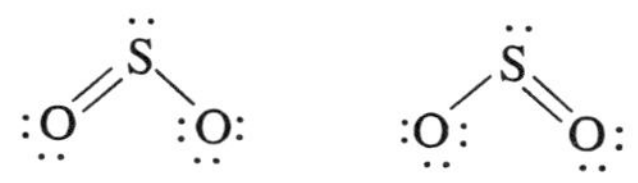

:Ö=S̈–Ö: ⟷ :Ö–S̈=Ö:

Since by the opening up of the bond, as indicated in the following structure,

:Ö–S̈–Ö:

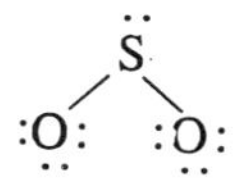

the sulphur atom is left with a vacant bonding orbital, it is not surprising that sulphur dioxide acts as an electron acceptor and forms molecular complexes with a variety of electron-donor molecules.

As might be predicted from its dielectric constant, sulphur dioxide is a better solvent for covalently bonded substances than for ionic substances.

The solubilities of some salts are given in Table 5.5. From this table it can be seen that only the alkali metal iodides are soluble to the extent of more than a mole per thousand grammes of solvent. The solubility of most of the others is only a few millimoles. All the tetramethylammonium halides are freely soluble in the solvent, because of their low lattice energies.

Bromine, chlorine, iodine monochloride, thionyl chloride, boron trichloride, carbon disulphide, phosphorous trichloride, arsenic trichloride, and phosphorus oxychloride are miscible in all proportions. Liquid sulphur dioxide is an excellent solvent for organic compounds. Amines, ethers, esters, alcohols, sulphides, mercaptans, and acids (both aliphatic and aromatic) are readily soluble. Aromatic hydrocarbons and alkenes dissolve readily, but paraffins have a low solubility. Water is soluble but not completely miscible with liquid sulphur dioxide.

Sulphur dioxide forms stable solvates with many alkali metal halides and other ionic compounds. The molar ratio of SO_2 to halide

Table 5.5 Solubilities of Alkali Metal, Ammonium, Silver, and Thallous Salts in Liquid SO_2

Ion	SO_3^{2-}	SO_4^{2-}	F^-	Cl^-	Br^-	I^-	SCN^-	CN^-	ClO_4^-	$CH_3 \cdot COO^-$
Li^+	—	1.55	23.0	2.82	6.0	1,490.0	—	—	—	3.48
Na^+	1.37	insol.	6.9	insol.	1.36	1,000.0	80.5	3.67	—	8.90
K^+	1.58	insol.	3.1	5.5	40.0	2,490.0	502.0	2.62	—	0.61
Rb^+	1.27	—	—	27.2	sol.	sol.	—	—	—	—
Cs^+	—	—	—	—	—	—	—	—	—	—
NH_4^+	2.67	5.07	Less than 27 at 50°C	1.67	6.0	580.0	6,160.0	—	2.14	141.0
Tl^+	4.96	0.417	insol.	0.292	0.60	1.81	0.915	0.522	0.43	285.0
Ag^+	insol.	insol.	insol.	20.07	0.159	0.68	0.845	1.42	—	1.02

Data refer to 0°C unless otherwise stated. Solubilities are in millimoles per 1,000 g of sulphur dioxide.

generally varies from one to four. Some results are summarized in Table 5.6. The mono-adducts are probably best regarded as halo-sulphinites, with structure

$$X{-}S\begin{smallmatrix} \nearrow O^- \\ \searrow O \end{smallmatrix}$$

by comparison with the halosulphonates. Their stability decreases in the order F > Cl > Br > I. The structure of the polysolvates has not been determined, but the iodides are here the most stable and coordination is probably through the sulphur atom to the anion.

Table 5.6 Solvate Formation with Alkali Metal and Tetramethylammonium Salts.

$LiI,2SO_2$	KF,SO_2	RbF,SO_2		Me_4NF,SO_2
$NaI,2SO_2$	Me_4NCl,SO_2			
$NaI,4SO_2$	$KBr,4SO_2$	$RbI,4SO_2$	$CsI,4SO_2$	$Me_4NCl,2SO_2$
$NaNCS,2SO_2$	$KI,4SO_2$	Me_4NBr,SO_2		
$NaCH_3{\cdot}CO_2,SO_2$	$KNCS,SO_2$			$Me_4NBr,2SO_2$
	$KCH_3{\cdot}CO_2,SO_2$	$RbCH_3{\cdot}CO_2,SO_2$		Me_4NI,SO_2
$(Me_4N)_2SO_4,3SO_2$				
				$(Me_4N)_2SO_4,6SO_2$

With covalent compounds many SO_2 solvates are formed. In most cases the molecule acts as an electron acceptor through sulphur and only in a few as an electron donor through oxygen. These latter seem to be restricted to only the most powerful Lewis acids and even then are not very stable. Examples are SO_2,BF_3, SO_2,SbF_5 (the most stable), SO_2, $SnBr_4$, and $SO_2,AlCl_3$.

Organic amines form stable mono-adducts with SO_2; these compounds are usually highly coloured and coordination appears to be through the nitrogen to the sulphur.

Molecular addition compounds with liquid sulphur dioxide are listed in Table 5.7.

Electrolytic Behaviour of Solutions in Liquid Sulphur Dioxide

Nearly all the experimental work in liquid sulphur dioxide is limited to uni-univalent electrolytes. These solutions are nothing like as good

Table 5.7 Molecular Addition Compounds of SO_2

Compound	*Adduct*
$(CH_3)_3N$	$(CH_3)_3N,SO_2$
$(C_2H_5)_3N$	$(C_2H_5)_3N,SO_2$
$C_6H_5\cdot N(CH_3)_2$	$C_6H_5\cdot N(CH_3)_2,SO_2$
$C_6H_5\cdot NH_2$	$C_6H_5\cdot NH_2,SO_2$
Pyridine	C_5H_5N,SO_2
p-$C_6H_4(NH_2)_2$	p-$C_6H_4(NH_2)_2,2SO_2$
$(C_2H_5)_3NO$	$(C_2H_5)_3NO,SO_2$
Ethylene oxide	$(CH_2)_2O,SO_2$
Dioxane	$O(CH_2\cdot CH_2)_2O,2SO_2$
$C_6H_5\cdot O\cdot CH_3$	$C_6H_5\cdot O\cdot CH_3,SO_2$
$CH_3\cdot COOH$	$CH_3\cdot COOH,SO_2$
$(C_2H_5)_2S$	$(C_2H_5)_2S,SO_2$

electrical conductors as liquid ammonia solutions or constant, equilibria exist in solution not only between ions and ion pairs (uncharged) but also in more concentrated solutions between ion pairs and ion triplets (charged). A plot of equivalent conductivity against concentration shows a minimum. At concentrations above about 0.1 M the conductivity is largely due to ion triplets; a minimum is found at about 0.1 M, and below this the equivalent conductivity increases with decreasing concentration, obeying Ostwald's dilution law. Triple ion formation can be disregarded at concentrations below 0.01 M, and at this concentration association constants for ion pair formation may be determined.

The data for a number of salts at 0°C are given in Table 5.8. The limiting conductivities of the salts in the table show that there must be considerable hydrodynamic transport of solvent associated with the lithium ion. With constant cation the association constants increase from $Cl^- < Br^- < I^-$, and with constant anion they increase from $Li^+ < Na^+ < K^+ < Me_4N^+ < Et_4N^+$.

Triphenylchloromethane and some of its chloro and methyl derivatives are moderately good conductors in liquid SO_2. The conductivity results may be explained in terms of the following reactions:

$$(C_6H_5)_3CCl \rightleftharpoons \underset{\text{ion pair}}{(C_6H_5)_3C^+Cl^-} \rightleftharpoons (C_6H_5)_3C^+ + Cl^-$$

Table 5.8 Conductivity and Ion Pair Association Data in Liquid Sulphur Dioxide

Ion pair	*Limiting molar conductance* ($cm^2\ ohm^{-1}\ mol^{-1}$)	K ($mol\ l^{-1} \times 10^4$)
LiBr	189	0.27
NaBr	265	0.48
KCl	243	0.74
KBr	249	1.43
KI	244	3.0
Me_4NCl	243	10.3
Me_4NBr	236	11.8
Me_4NI	234	13.9
Me_4NClO_4	218	8.4
Me_4NBF_4	215	7.9
Et_4NBr	215	21
Et_4NI	197	39

with an equilibrium constant:

$$K = \frac{[(C_6H_5)_3C^+][Cl^-]}{[(C_6H_5)_3CCl] + \underset{\text{ion pair}}{[(C_6H_5)_3C^+Cl^-]}}$$

Solutions of trisubstituted hydronium salts in the solvent are also fairly good electrical conductors. Salts like $(CH_3)_3O^+BF_4^-$, $(C_2H_5)_3O^+BF_4^-$, and $(C_2H_5)_3O^+SbCl_6^-$ have conductivities of about the same order of magnitude as solutions of KI. It is also worth noting that, though neither water nor hydrogen bromide by themselves are electrical conductors in sulphur dioxide, a mixture of both gives a conducting solution in the solvent, presumably due to $H_3O^+Br^-$.

Reactions in Liquid Sulphur Dioxide

Autoionization and Acid-Base Reactions. The electrical conductivity of pure liquid sulphur dioxide (4×10^{-8} ohm^{-1} cm^{-1} at $-10\,°C$) is commonly interpreted in terms of self-ionization in accordance with the following equation:

$$2SO_2 \rightleftharpoons SO^{++} + SO_3^{=}$$

which may be considered as analogous to the corresponding self-ionization reactions for water and liquid ammonia:

$$2H_2O \rightleftharpoons H_3O^+ + OH^-$$
$$2NH_3 \rightleftharpoons NH_4^+ + NH_2^-$$

According to this formal analogy, the SO^{++} ion would be the analogue of the hydronium and ammonium ions, and the sulphite ion, $SO_3^=$, would be analogous to the hydroxide and amide ions.

Thus, typical bases in the sulphur dioxide systems would include those compounds which contain or make available sulphite ion. Thus, the alkali metal sulphites should behave as typical bases in liquid sulphur dioxide. The actual species present is probably the pyrosulphite ion, $S_2O_5^=$, formed by the solvation reaction

$$SO_3^= + SO_2 \rightarrow S_2O_5^=$$

In the reactions which follow we shall write it as simply sulphite, $SO_3^=$.

Similarly, it would be expected that compounds which contain or make available the SO^{++} ion would be typical acids in liquid sulphur dioxide. Thus, thionyl compounds, such as $SOCl_2$ or $SOBr_2$, would behave as acids. It is not necessary that these thionyl compounds be actually appreciably ionized into SO^{++} ions but only that they are capable of reacting in such a way as to make this ion available in a chemical reaction.

Several typical acid-base reactions which have been carried out in liquid sulphur dioxide include the following:

$$Cs_2SO_3 + SOCl_2 \rightarrow 2CsCl + 2SO_2$$
$$K_2SO_3 + SO(SCN)_2 \rightarrow 2KSCN + 2SO_2$$
$$[(CH_3)_4N]_2SO_3 + SOBr_2 \rightarrow 2[(CH_3)_4N]Br + 2SO_2$$

Whereas, in analogy with reactions between bases and acids in aqueous solution, it is tempting to assume that the reactions between sulphites and thionyl compounds in sulphur dioxide occur by ionic mechanisms and may be represented by the equation

$$SO_3^= + SO^{++} \rightleftharpoons 2SO_2$$

it is not necessary to make this assumption. In fact, studies of the kinetics of exchange of radioisotopes of sulphur between thionyl halides and sulphur dioxide, as well as between tetramethylammonium pyrosulphite and sulphur dioxide, make it quite clear that the free ion

SO^{++} plays no important role in the mechanism of reactions in liquid sulphur dioxide. Mechanisms involving the transfer of $O^{=}$ ions have been suggested for the explanation of acid-base phenomena in liquid sulphur dioxide.

Experimental evidence supports the partial ionization of thionyl halides as, for example, the following equation indicates:

$$SOCl_2 \rightleftharpoons SOCl^+ + Cl^-$$

Solvolysis reactions. Zinc diethyl reacts even at dry ice temperatures to give diethyl sulphoxide and zinc oxide:

$$Zn(C_2H_5) + SO_2 \rightarrow ZnO + (C_2H_5)_2SO$$

Ammonium acetate is also solvolysed in liquid SO_2:

$$2NH_4(CH_3\cdot CO_2) + 2SO_2 \rightarrow (NH_4)_2SO_3 + (CH_3\cdot CO_2)_2SO$$

$$(CH_3\cdot CO_2)_2SO \rightarrow (CH_3\cdot CO)_2O + SO_2$$

There is some dispute as to whether solutions of alkali metal bromides and iodides are slowly solvolysed in the solvent.

The following sequence of reactions have been reported:

$$8KBr + 8SO_2 \rightarrow 4K_2SO_3 + 4SOBr_2$$

$$4SOBr_2 \rightarrow 2SO_2 + S_2Br_2 + 3Br_2$$

$$\underline{4K_2SO_3 + 2Br_2 \rightarrow 2K_2SO_4\downarrow + 4KBr + 2SO_2}$$

$$4KBr + 4SO_2 \rightarrow 2K_2SO_4\downarrow + S_2Br_2 + Br_2$$

Presumably the same reactions take place with KI, except that S_2I_2 breaks down to sulphur and iodine:

$$4KI + 4SO_2 = 2K_2SO_4\downarrow + 2S\downarrow + 2I_2$$

The behaviour of the binary halides in liquid sulphur dioxide varies. Halides of Group IV do not react. Phosphorus pentachloride and pentabromide are readily solvolysed, even at low temperatures:

$$PCl_5 + SO_2 \rightarrow POCl_3 + SOCl_2$$

$$PBr_5 + SO_2 \rightarrow POBr_3 + SOBr_2$$

Further solvolysis does not occur. This is because, though the reaction to the oxyhalide is exothermic, the further solvolysis:

$$2POCl_3 + 3SO_2 \rightarrow (P_2O_5) + 3SOCl_2$$

would represent an endothermic reaction. Other solvolytic reactions which do occur are:

$$NbCl_5 + SO_2 \rightarrow NbOCl_3 + SOCl_2$$
$$WCl_6 + SO_2 \rightarrow WOCl_4 + SOCl_2$$
$$UCl_6 + 2SO_2 \rightarrow UO_2Cl_2 + 2SOCl_2$$

Metathetical reactions. A large number of reactions, which have been called 'neutralization reactions' or 'acid-base reactions', are probably better described as metathetical reactions. Thus, sulphites react with thionyl halides in the solvent to produce chlorides and sulphur dioxide:

$$Cs_2SO_3 + SOCl_2 \rightarrow 2CsCl + 2SO_2$$
$$[Me_4N]_2SO_3 + SOBr_2 \rightarrow 2Me_4NBr + 2SO_2$$

Acetates also react:

$$2Ag(CH_3{\cdot}CO_2) + SOCl_2 \rightarrow 2AgCl\downarrow + SO(CH_3{\cdot}CO_2)_2$$
$$2NH_4(CH_3{\cdot}CO_2) + SOCl_2 \rightarrow 2NH_4Cl\downarrow + SO(CH_3{\cdot}CO_2)_2$$

$SO(CH_3{\cdot}CO_2)_2$ has not been isolated; it appears to break up into acetic anhydride and sulphur dioxide. However, $(C_6H_5{\cdot}CH_2{\cdot}CO_2)_2SO$ and $(ClCH_2{\cdot}CO_2)_2SO$ have been isolated from similar reactions. Ammonium thiocyanate also reacts with thionyl chloride:

$$2NH_4SCN + SOCl_2 \rightarrow 2NH_4Cl\downarrow + SO(SCN)_2$$

The $SO(SCN)_2$ is stable in dilute solution and can be titrated conductimetrically with K_2SO_3:

$$K_2SO_3 + SO(SCN)_2 \rightarrow 2KSCN + 2SO_2$$

Amphoteric reactions. A number of reactions in liquid sulphur dioxide appear to be analogous to the behaviour of amphoteric substances in water. When a solution of tetramethylammonium sulphite is added to a solution of aluminium chloride in liquid sulphur dioxide, a voluminous white precipitate of aluminium sulphite is formed:

$$2AlCl_3 + 3(Me_4N)_2SO_3 \rightarrow Al_2(SO_3)_3\downarrow + 6Me_4NCl$$

If an excess of tetramethylammonium sulphite is then added the precipitate redissolves, presumably with formation of a complex anion:

$$Al_2(SO_3)_3 + 3(Me_4N)_2SO_3 \rightarrow 2(Me_4N)_3Al(SO_3)_3$$

If thionyl chloride is now added to this solution the aluminium sulphite is reprecipitated:

$$2(Me_4N)_3Al(SO_3)_3 + 3SOCl_2 \rightarrow Al_2(SO_3)_3\downarrow + 6Me_4NCl + 6SO_4$$

Excess of thionyl chloride not redissolve the aluminium sulphite, however, a deviation from the analogy with aqueous solutions.

Similar reactions are observed with gallium trichloride. When tetramethylammonium sulphite is added to a solution of gallium trichloride a precipitate is formed:

$$2GaCl_3 + 3(Me_4N)_2SO_3 \rightarrow 6Me_4NCl + Ga_2(SO_3)_3\downarrow$$

When excess of the sulphite is added, the precipitate redissolves. Again, stannic chloride solutions behave similarly; sulphite precipitates stannic sulphite, and this dissolves in excess of sulphite to give a solution of orthosulphitostannate, $(Me_4N)_4Sn(SO_3)_4$. A solution of PCl_3 in liquid sulphur dioxide, which is itself stable, gives, when a solution of tetramethylammonium sulphite is added, a flocculent precipitate of phosphorus trioxide:

$$2PCl_3 + 3(Me_4N)_2SO_3 \rightarrow P_2O_3 + 3SO_2 + 6Me_4NCl$$

On addition of further sulphite the precipitate redissolves, and from this solution the compound $Me_4NPO_2SO_2$ has been isolated:

$$P_2O_3 + (Me_4N)_2SO_3 + SO_2 \rightarrow 2(Me_4N)PO_2SO_2$$

This reaction has been followed conductimetrically and two end-points have been observed, the first at three moles of sulphite to two of PCl_3 and the second at four to two.

Metals which have amphoteric hydroxides generally dissolve in aqueous alkali with the liberation of hydrogen. Most of these, including beryllium, aluminium, gallium, antimony, and lead, give no reaction with tetramethylammonium sulphite in liquid sulphur dioxide. However, tin does react. With excess of tetramethylammonium sulphite the following reaction takes place:

$$Sn + 2(Me_4N)_2SO_3 + 3SO_2 \rightarrow (Me_4N)_2Sn(SO_3)_3 + (Me_4N)_2S_2O_3$$

the sulphur dioxide being reduced to thiosulphate.

Complex formation. Added iodine increase the conductivity of potassium iodide and rubidium iodide in liquid SO_2, and the solubility of iodine itself is greatly increased by the addition of potassium iodide or rubidium iodide. These effects are a maximum at an iodine:iodide ratio of 1:1 and are due to the formation of tri-iodide ion:

$$KI + I_2 \rightarrow KI_2$$

Similarly the solubilities of cadium iodide and mercuric iodide are increased by the addition of potassium iodide and rubidium iodide to the solvent, owing to the formation of complex ions:

$$HgI_2 + 2KI \rightarrow K_2HgI_4$$

Addition of $SbCl_3$ to a solution of KCl gives a precipitate of K_3SbCl_6:

$$3KCl + SbCl_3 \rightarrow K_3SbCl_6 \downarrow$$

Addition of excess of antimony pentachloride causes this complex to dissolve with decomposition of the complex ion and formation of a hexachloroantimonate:

$$K_3SbCl_6 + 3SbCl_5 \rightarrow 3KSbCl_6 + SbCl_3$$

The preparation of a number of hexachloroantimonates in liquid SO_2 has been followed conductimetrically. When NOCl, $CH_3 \cdot COCl$, and $C_6H_6 \cdot COCl$ are titrated against $SbCl_5$ in SO_2, the conductimetric titrations show a sharp break at a ratio of 1:1, and the compounds NO^+SbCl_6, $CH_3 \cdot CO^+SbCl_6^-$ and $C_6H_5 \cdot CO^+SbCl_6^-$ can be isolated.

Oxidation-reduction reactions. In many oxidation-reduction reactions studied in SO_2, the solvent acts merely as an inert carrier. Tetramethylammonium sulphite is rapidly oxidized to the sulphate by iodine. A solution of ferric chloride will quantitatively oxidize KI to iodine:

$$2FeCl_3 + 2KI \rightarrow 2FeCl_2 + 2KCl + I_2$$

and so will a solution of antimony pentachloride:

$$6KI + 3SbCl_5 \rightarrow 3I_2 + 6KCl + 3SbCl_3$$

The reaction is complicated by the reaction of $SbCl_3$ with KCl and then the further reaction of excess of $SbCl_5$ with the precipitated K_3SbCl_6. In a conductimetric titration, in addition to a break at a mole ratio of $SbCl_5$ to KI of 1:2, there is a break at a mole ratio of 3:2 corresponding to:

$$2KI + 3SbCl_5 \rightarrow I_2 + 2KSbCl_6 + SbCl_3.$$

Nitrosyl compounds can act as oxidizing agents in liquid SO_2. Thus, with tetramethylammonium iodide:

$$2NOX + 2I^- \rightarrow 2NO + I_2 + 2X^- \quad (X^- = Cl^- \text{ or } BF_4^-)$$

Ethyl nitrite acts in a similar manner:

$$2C_2H_5{\cdot}O{\cdot}NO + 2SO_2 + 2I^- \rightarrow 2NO + I_2 + 2SO_2{\cdot}OC_2H_5^-$$

With azides the reactions are:

$$NOX + N_3^- \rightarrow N_2O + N_2 + X^- \quad (X^- = Cl^- \text{ or } BF_4^-)$$

and

$$C_2H_5{\cdot}O{\cdot}NO + N_3^- + SO_2 \rightarrow N_2O + N_2 + SO_2{\cdot}OC_2H_5^-$$

Reactions of organic compounds. Because many organic compounds are very soluble in liquid SO_2 and because it is itself inert to many of them, a range of synthetic organic reactions has been carried out in this solvent. These include sulphonations by sulphur trioxide or by chlorosulphuric acid:

$$C_6H_6 + SO_3 \rightarrow C_6H_5{\cdot}SO_3H$$
$$C_6H_6 + HSO_3Cl \rightarrow C_6H_5{\cdot}SO_3H + HCl$$

Sulphur dioxide has also been used as a solvent for Friedel-Crafts reactions because of the high solubility of aluminium chloride in it:

$$C_6H_6 + (CH_3)_3CCl \xrightarrow[SO_2]{AlCl_3} C_6H_5{\cdot}C(CH_3)_3 + HCl$$

$$C_6H_6 + C_6H_5{\cdot}C(=O)Cl \xrightarrow[SO_2]{AlCl_3} (C_6H_5)_2CO + HCl$$

It has also been used as a solvent for the addition of bromine to unsaturated compounds:

$$C_6H_5{\cdot}CH{:}CH_2 + Br_2 \rightarrow C_6H_5{\cdot}CHBr{\cdot}CH_2Br$$

LIQUID HALIDES AS IONIZING SOLVENTS

The behaviour of some halides, usually considered to be covalent, which function as ionizing solvents is now considered. They are AsF_3,

$AsCl_3$, $AsBr_3$, and $SbCl_3$, and the interhalogen compounds ICl and BrF_3.

Arsenic Trihalides

Of the arsenic trihalides only the trichloride has been studied in some detail.

A comparison of physical properties is given in Table 5.9.

Table 5.9 Physical Properties of the Arsenic Trihalides

	AsF_3	$AsCl_3$	$AsBr_3$
Melting point (°C)	−6	−13	35
Boiling point (°C)	63	130	220
Liquid range (°C)	69	143	185
Viscosity (centipoise)	8.57 (20°C)	1.225 (20°C)	5.41 (35°C) 4.44 (48°C)
Density (g cm^{-3})	2.45 (20°C)	2.16 (20°C)	3.33 (50°C)
Dielectric constant	5.7 (−6°C)	12.6 (17°C)	8.8 (35°C)
Specific conductivity (ohm cm^{-1})	2.4×10^{-5}	1.4×10^{-7}	1.6×10^{-7}

Arsenic trichloride: As a solvent, the suggested self-ionization is:

$$2AsCl_3 \rightleftharpoons AsCl_2^+ + AsCl_4^-$$

and similar self-ionizations would probably also apply to the other trihalides. The dielectric constant is fairly small; consequently, the solubilities of the alkali metal chlorides and ammonium chloride are very low. However, the tetra-alkylammonium halides are readily soluble. $NbCl_5$ and $TaCl_5$ are not very soluble; but $AlCl_3$, $SnCl_4$, VCl_4, and $FeCl_3$ are very soluble. Various non-metals such as sulphur, phosphorus, and iodine also dissolve readily, but the nature of these solutions is unknown. Metals, metal oxides, and oxy-salts, such as sulphates and nitrates, are either insoluble or have very low solubilities. A wide range of solvates is formed by the solvent, mostly of halides.

Chemical reactions in arsenic trichloride: Solutions of both tetramethylammonium chloride and antimony pentachloride have high conductivities in $AsCl_3$. A conductimetric titration of these solutions shows a break at a 1:1 end-point.

The salt Me_4NSbCl_6 can be isolated from the solution. The reaction is probably:

$$Me_4N^+AsCl_4^- + AsCl_2^+SbCl_6^- = Me_4N^+SbCl_6^- + 2AsCl_3$$

since the antimony pentachloride solution has a high conductivity.

Tellurium tetrachloride dissolves in $AsCl_3$ to give a conducting solution which, in a conductimetric titration against Me_4NCl, gives end-points at an acid-base ratio of 1:1 and 1:2. From the 1:1 solution a solid of composition $Me_4NCl,TeCl_4,AsCl_3$ can be isolated, which may be formulated as $Me_4N^+AsCl_2^+TeCl_6^{2-}$. From the 1:2 solution, solid $(Me_2N^+)_2TeCl_6^{2-}$ can be obtained. When $TeCl_4$ is titrated with $SbCl_5$ in $AsCl_3$, breaks are found at molar ratios of 2:1 and 1:1 $TeCl_4$ to $SbCl_5$. From the 2:1 solution a solid of composition $2TeCl_4,SbCl_5,AsCl_2$ has been isolated. This can be formulated as $(TeCl_3^+)_2AsCl_4^-SbCl_6^-$. From the 1:1 solution a solid of composition $TeCl_4,SbCl_5$, which is probably $TeCl_3^+SbCl_6^-$, can be isolated.

Pyridine solutions in arsenic trichloride are good conductors of electricity and probably function by liberating chloride ion:

$$C_5H_5N + 2AsCl_3 \rightarrow C_5H_5N\cdot AsCl_2^+ + AsCl_4^-$$

Such solutions can be titrated against $SnCl_4$ and VCl_4. In both cases the conductimetric titration curves show a break at an end-point of 1:2 and solids of composition $2C_5H_5N,2AsCl_3,SnCl_4$ and $2C_5H_5N,2AsCl_3,VCl_4$, which are presumably $(C_5H_5N\cdot AsCl_2^+)_2SnCl_6^{2-}$ and $(C_5H_5\text{-}N\cdot AsCl_2^+)_2VCl_6^{2-}$.

Arsenic trifluoride: It has a fairly high specific conductivity, but this may be due to impurities. In spite of its low dielectric constant, potassium fluoride dissolves readily in the solvent, as do rubidium fluoride and caesium fluoride. From these solutions the compounds $KAsF_4$, $RbAsF_4$ can be isolated, and it is the formation of the complex ion, AsF_4^-, and of probably higher complexes in solution that explains the solubility of the fluorides.

Antimony pentafluoride also dissolves in the solvent to give a conducting solution, from which a compound of composition AsF_3,SbF_5, which can be formulated as $AsF_2^+SbF_6^-$, can be isolated.

Conductimetric titrations of solutions of potassium fluoride and antimony pentafluoride give a 1:1 end-point and also produce solid $KSbF_6$:

$$AsF_2^+SbF_6^- + K^+AsF_4^- \rightarrow KSbF_6 \downarrow + 2AsF_3$$

Chemical reactions in arsenic trifluoride: When chlorine is passed into AsF_3 at 0°C a white crystalline substance, $AsCl_4^+AsF_6^-$, is formed. Many chlorine-containing compounds react with the solvent. Thus $SbCl_5$ reacts as follows:

$$3SbCl_5 + AsF_3 \rightarrow 3SbCl_4^+F^- + AsCl_3$$

N.M.R. studies of arsenic trifluoride. ^{19}F studies of AsF_3 and of solutions of alkali metal fluorides show that the later have only one resonance line, shifted slightly from that of AsF_3. This indicates that the solvated fluoride ion, or the AsF_4^- ion, exchanges fluorines sufficiently rapidly with AsF_3 to give only a single n.m.r. resonance. Similarly, the ^{19}F spectrum of an arsenic trifluoride-antimony pentafluoride mixture shows only one resonance peak, in a position between the resonances of the two pure compounds. This means that rapid exchange between AsF_3 and SbF_5 via fluoride bridges must occur:

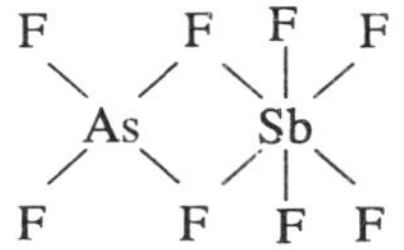

```
F       F  F  F
 \     / \ | /
   As      Sb
 /     \ / | \
F       F  F  F
```

Arsenic tribromide: The alkali, alkaline earth, and divalent transition metal bromides and salts of oxy-acids and metal oxides do not appear to have an appreciable solubility. $HgBr_2$, $InBr_3$, $TeBr_4$, and $BiBr_3$ are all moderately soluble. Quaternary ammonium bromides and BBr_3, $AlBr_3$, $GaBr_3$, $SnBr_4$, $TiBr_4$, PBr_5, $SbBr_3$, and $SeBr_4$ are readily soluble. Many organic compounds, including aromatic hydrocarbons, alcohols, ketones, aldehydes, esters, and amines, are freely soluble in the solvent.

Solvates of the type $R_4NBr,AsBr_3$ are formed with quaternary ammonium bromides, and with other organic bases, *B*, solvates of the type *B*,HBr,$AsBr_3$ are found. Both types should probably be formulated as containing the $AsBr_4^-$ ion. Cryoscopic measurements show that with anhydro bases like triethylamine and pyridine an equilibrium occurs:

$$(C_2H_5)_3N + 2AsBr_3 \rightleftharpoons (C_2H_5)_3N \cdot AsBr_2^+ + AsBr_4^-$$

Addition of silver salts, such as silver perchlorate, at 50°C leads to the precipitation of AgBr and a solution of high conductivity;

$$AsBr_3 + AgClO_4 = AsBr_2^+ClO_4^- + AgBr \downarrow$$

It is, however, not possible to isolate $AsBr_2^+ClO_4^-$ as a solid compound.

The Lewis acids $AlBr_3$, $GaBr_3$, $InBr_3$, BBr_3, $SnBr_4$, $HgBr_2$, and $TeBr_4$ may be employed in conductimetric titration against tetraalkylammonium bromide in the solution and a variety of salts may be isolated from this solution.

Antimony Trichloride

Some of the physical properties of antimony trichloride are listed in Table 5.10.

Table 5.10 Physical Properties of $SbCl_3$

Melting point (°C)	73
Boiling point (°C)	219-223
Liquid range (°C)	150
Viscosity (centipoise)	3.3 (95°C)
Density (g cm^{-3})	2.44 (178°C)
Dielectric constant	33.0 (75°C)
Specific conductivity (ohm^{-1} cm^{-1})	0.85×10^{-6} (95°C)
Trouton's constant	~ 21
Enthalpy of vaporization (kcal mol^{-1})	10–36
Molar cryoscopic constant (deg mol^{-1} kg)	15.6 ± 0.2

Unlike their behaviour in arsenic trichloride, KCl, RbCl, CsCl, and NH_4Cl are readily soluble in antimony trichloride, as are the quaternary ammonium chlorides. $HgCl_2$, $AlCl_3$, $SbCl_5$, and $TeCl_4$ are also readily soluble. However, LiCl, NaCl, $SnCl_2$, $BiCl_3$, and $FeCl_3$ have only at most a slight solubility. With the exception of their tetraalkylammonium salts, sulphates and perchlorates are insoluble. Many organic compounds such as aromatic hydrocarbons, organic halides, and oxygen- and sulphur-containing compounds are soluble, and the solubility of their infrared O—H and N—H vibration frequencies. A wide variety of solvates with inorganic chlorides is formed; some of these are $KSbCl_4$, K_2SbCl_5, KSb_2Cl_7, $K_3Sb_2Cl_9$ $Ba(SbCl_4)_2$, and $BaSbCl_5$.

Chemical reactions in antimony trichloride: Silver perchlorate dissolves readily in $SbCl_3$ to give a precipitate of silver chloride and a conducting solution. The reaction is probably:

$$AgClO_4 + SbCl_3 \rightarrow AgCl \downarrow + SbCl_2^+ClO_4^-$$

Similarly, $AlCl_3$ and $SbCl_5$ give solutions of high conductivity, e.g.:

$$SbCl_5 + SbCl_3 \rightarrow SbCl_2^+ + SbCl_6^-$$

Conductimetric titrations between such solutions and solutions of tetramethylammonium chloride in the solvent give 1:1 end-points:

$$Me_2N^+SbCl_4^- + SbCl_2^+ClO_4 \rightarrow Me_4N^+ClO_4^- + 2SbCl_3$$

Solutions of many organic chlorides in $SbCl_3$ are quite good conductors, and it has been shown that these compounds ionize mainly by the reaction:

$$2RCl + SbCl_3 \rightleftharpoons R_2Cl^+ + SbCl_4^-$$

in concentrated solutions.

Iodine Monochloride

Some of the physical properties of iodine monochloride are given in Table 5.11. Potassium chloride and ammonium chloride are moderately soluble in the solvent, So too are RbCl, CsCl, KBr, KI, $AlCl_3$, $AlBr_3$, PCl_5, pyridine, acetamide, and benzamide. All of these compounds yield conducting solutions. LiCl, NaCl, AgCl, and $BaCl_2$ are sparingly soluble. Silicon tetrachloride, titanium tetrachloride and niobium pentachloride dissolve but do not enhance the conductivity. Antimony pentachloride dissolves and increases the conductivity. Thionyl chloride is miscible in all proportions, but simply acts as a

Table 5.11 Physical Properties of Iodine Monochloride

Melting point (°C)	27.2 (α), 13.9 (β)
Boiling point (°C)	100
Liquid range (°C)	73
Density (g cm^{-3})	3.24 (34°C)
Specific conductivity (ohm^{-1} cm^{-1})	4.6×10^{-3} (35°C)
Trouton's constant	26.7
Enthalpy of vaporization (kcal mol^{-1})	10.0

diluent and its addition decreases the conductivity of the mixture. However, the mixture is in some ways a much better solvent than either thionyl chloride or iodine monochloride by itself. Potassium chloride is much more soluble in the mixture than in either of the pure components. This is presumably due to the ability of thionyl chloride to coordinate the cation and iodine monochloride to coordinate the anion.

The importance of self-ionization in the solvent is indicated by the high specific conductivity. Interpretation is complicated by the dissociation into chlorine and iodine. The degree of dissociation is 0.4% at 25°C and 1.1% at 100°C. The self-ionization is probably best represented by:

$$3ICl \rightleftharpoons I_2Cl^+ + ICl_2^-$$

The existence of ICl_2^- in $PCl_4^+ICl_2^-$ and in $KICl_2$ and $RbICl_2$ is well established by X-ray studies. The phase diagrams of the systems $AlCl_3$–ICl and $SbCl_5$–ICl show the existence of compounds $AlCl_3$, 2ICl and $SbCl_5$, 2ICl, which may be formulated as $I_2Cl^+AlCl_4^-$ and $I_2Cl^+SbCl_6^-$. Limiting conductivities, Λ_0, at 35°C of 32 ohm^{-1} cm^2 mol^{-1} for KCl and 26 ohm^{-1} cm^2 mol^{-1} for NH_4Cl are reported. Conductimetric titrations of RbCl against $SbCl_5$ and KCl against $NbCl_5$ show a 1:1 break, and conductimetric titrations of NH_4Cl against $SnCl_4$ show a base:acid break of 2:1.

Bromine Trifluoride

Some of the more important physical properties of bromine trifluoride are listed in Table 5.12. Bromine trifluoride fluorinates virtually everything which dissolves in it, so a discussion of solubilities

Table 5.12 Physical Properties of Bromine Trifluoride

Property	Value
Melting point (°C)	9
Boiling point (°C)	126
Liquid range (°C)	117
Viscosity (centipoise)	2.22(25°C)
Density (g cm^{-3})	2.8 (25°C)
Specific conductivity (ohm^{-1} cm^{-1})	8.0×10^{-3} (25°C)
Trouton's constant	25.6
Enthalpy of vaporization (kcal mol^{-1})	10.2

is restricted to soluble inorganic fluorides. The alkali metal fluorides, silver fluoride and barium fluorides, are readily soluble. The fluorides AuF_3, BF_3, TiF_4, SiF_4, GeF_4, VF_4, NbF_5, TaF_5, PF_5, AsF_5, SbF_5, PtF_4, and RuF_5 are also readily soluble. All the above are conductors in BrF_3. Any of the alkali metal or silver or barium fluorides reacts with the Lewis fluorides listed to give a complex halide.

A number of adducts are formed both by the alkali metal fluorides and by Lewis acid fluorides. Thus the alkali metal fluorides form the adducts MF, BrF_3, which can be formulated as $M^+BrF_4^-$. Stannic fluoride forms the adduct $2BrF_3,SnF_4$ and antimony pentafluoride the adduct BrF_3, SbF_5, which may be formulated respectively as $(BrF_2^+)_2SnF_6^{2-}$ and $BrF_2^+SbF_6^-$.

Chemical reactions in bromine trifluoride: Apart from the neutralization reactions already discussed a very large number of complex fluoro-compounds can be prepared in the solvent. Auric fluoride was prepared for the first time by the reaction of metallic gold with BrF_3:

$$Au + BrF_3 \rightarrow BrF_2^+AuF_4^- \rightarrow AuF_3$$

When a mixture of silver and gold is used the compound $AgAuF_4$ is formed.

A mixture of ruthenium metal and potassium chloride is converted into potassium hexafluororuthenate:

$$Ru + KCl \xrightarrow{BrF_3} KRuF_6$$

Nitrosonium compounds may be produced from nitrosyl chloride and Lewis acid fluorides:

$$NOCl + SnF_4 \xrightarrow{BrF_3} (NO)_2SnF_6$$

OXYHALIDE SOLVENTS

Now, we will consider the oxyhalides NOCl and $POCl_3$ as solvents. Apart from the many useful preparations that can be carried out in them, their chief interest lies in their mode of action.

Nitrosyl Chloride

Some of the important physical properties of nitrosyl chloride are listed in Table 5.13. On the basis of its reasonably high dielectric

Table 5.13 Physical Properties of Nitrosyl Chloride

Melting point (°C)	−64.5
Boiling point (°C)	−5.5
Liquid range (°C)	59
Viscosity (centipoise)	0.586 (−27°C)
Density (centipoise)	1.59 (−6°C)
Dielectric constant	19.7 (10°C); 22.5 (−27°C)
Specific conductivity (ohm^{-1} cm^{-1})	2.88×10^{-6} (−20°C)
Dipole moment (Debye)	1.83

constant nitrosyl chloride would be expected to be a fairly good ionizing solvent. However, materials such as potassium chloride are insoluble in the solvent and it is necessary to go to the tetra-alkylammonium halides to achieve solubility. Nitrosyl chloride has, however, remarkable solvent properties for nitrosonium salts.

Compounds such as $(NO)AlCl_4$, $(NO)FeCl_4$, and $(NO)SbCl_6$ are readily soluble and are strong electrolytes. However the 2:1 electrolytes $(NO)_2SnCl_6$ and $(NO)_2TiCl_6$ are not soluble.

Conductimetric titrations can also be carried out in the solvent.

Isotopic studies using radioactive chlorine, ^{36}Cl, have shown that rapid exchange of chlorine takes place between $AlCl_3$, $GaCl_3$, $InCl_3$, $TeCl_3$, $FeCl_3$, and $SbCl_5$ and the solvent. All chlorine atoms in these materials appear to be equivalent. Some exchange of ^{36}Cl between $ZnCl_2$, $CdCl_2$, and $HgCl_2$, which are not significantly soluble, and nitrosyl chloride also takes place fairly rapidly. This is presumed to take place as a result of the formation of the 1:1 complex, which is insoluble. This is followed a slower heterogenous exchange of chlorine between the complex and the solvent. No exchange was found with sodium chloride and potassium chloride and the solvent. This must be due to their insolubility.

Not many studies of chemical reactions in the solvent have been made, but the reactions of the silver phosphates are worthy of note:

$$4AgPO_4 + 12NOCl \rightarrow (NO)_2P_4O_{11} + 5N_2O_3 + 12AgCl$$

$$3Ag_4P_2O_7 + 12NOCl \rightarrow (NO)_4P_6O_{17} + 4N_2O_3 + 12AgCl$$

$$Ag_3P_3O_9 + 3\ NOCl \rightarrow (NO)P_3O_8 + N_2O_3 + 3AgCl$$

$$Ag_4P_4O_{12} + 4NOCl \rightarrow (NO)_2P_4O_{11} + N_2O_3 + 4AgCl$$

The reactions produce additional P—O—P bond formation, rather surprisingly in the case of the silver tri- and tetra-metaphosphates.

Phosphoryl Chloride

Some important physical properties of phosphoryl chloride are listed in Table 5.14. It is worth noting the comparatively long liquid range of $POCl_3$ and its much higher viscosity than NOCl, in spite of its lower dielectric constant. This suggests that it is probably more associated than NOCl.

Table 5.14 Physical Properties of Phosphoryl Chloride

Melting point (°C)	1
Boiling point (°C)	108
Liquid range (°C)	107
Viscosity (centipoise)	1.15 (25°C)
Density (g cm^{-3})	1.71 (0°)
Dielectric constant	13.9 (22°C)
Specific conductivity (ohm^{-1} cm^{-1})	$< 2 \times 10^{-8}$
Trouton's constant	24.9
Enthalpy of vaporization (kcal mol^{-1})	9.5

In general the solubilities of the alkali halides in phosphoryl chloride are low, but are a great deal higher than in nitrosyl chloride, again in spite of the lower dielectric constant. The alkaline earth halides, silver chloride, mercurous chloride, and thallous chloride are insoluble.

Quaternary ammonium salts are very soluble. Compounds such as $SiCl_4$, $SiBr_4$, and $SnBr_4$ dissolve readily and are molecular in the solvent. PCl_5, $AsCl_3$, $BiCl_3$, ICl_3, SCl_4, $PtCl_4$, PBr_5, $BiBr_3$, and BiI_3 are readily soluble, but cryoscopic measurements indicate that they are dissociated. BCl_3, $AlCl_3$, $SbCl_5$, $SnCl_4$, $TiCl_4$, $TeCl_4$, and BBr_3 all are readily soluble but form adducts with the solvent.

The dielectric constant of $POCl_3$ is sufficiently low for soluble ionic compounds to show a molar conductivity minimum in the solvent, just as in SO_2. However, at concentrations below about 5×10^{-3} M, triple ions are in negligible concentration and the data can be interpreted in terms of an ion pair-free ion equilibrium,

$$[M^+X^-] \rightleftharpoons M^+ + X^-$$